CHICAGO PUBLIC LIBRARY
WOODSON REGIONAL
9525 S. HALSTED ST.   60628

POLITICAL ECONOMY

POLITICAL ECONOMY

ALLAN H. MELTZER
ALEX CUKIERMAN
SCOTT F. RICHARD

New York   Oxford
OXFORD UNIVERSITY PRESS
1991

Oxford University Press

Oxford   New York   Toronto
Delhi   Bombay   Calcutta   Madras   Karachi
Petaling Jaya   Singapore   Hong Kong   Tokyo
Nairobi   Dar es Salaam   Cape Town
Melbourne   Auckland

and associated companies in
Berlin   Ibadan

Copyright © 1991 by Oxford University Press, Inc.

Published by Oxford University Press, Inc.,
200 Madison Avenue, New York, New York 10016

Oxford is a registered trademark of Oxford University Press

All rights reserved. No part of this publication may be reproduced,
stored in a retrieval system, or transmitted, in any form or by any means,
electronic, mechanical, photocopying, recording, or otherwise,
without the prior permission of Oxford University Press.

Library of Congress Cataloging-in-Publication Data
Meltzer, Allan H.
Political economy / Allan H. Meltzer, Alex Cukierman, Scott F. Richard.
p. cm.   Includes bibliographical references and index.
ISBN 0-19-505624-8
1. Economics—Political aspects. 2. Decision-making.
3. Economic policy.   4. Policy sciences.
I. Cukierman, Alex.   II. Richard, Scott F.   III. Title.
HB73.M45   1991   320'.6—dc20   90-43235
ISBN 0-19-505624-8

1 3 5 7 9 8 6 4 2

Printed in the United States of America
on acid-free paper

To: Karl Brunner, scholar and friend.
He helped many of us to see the power of economics
for understanding individuals
and social organization.

Preface

The chapters in this volume started with a heated discussion following a seminar at the Graduate School of Industrial Administration. The seminar concerned the future of the American corporation and the long-term effects of government policies on corporations and stockholder wealth. The discussion soon broadened into the role of government in a democratic, free-market society. Two of us started to research the subject. As with most research, one idea leads to another. This book is the result.

Our interest in political economy had been stimulated, and our thinking aided, by several years of participation at the Interlaken Seminar on Analysis and Ideology that Karl Brunner organized beginning in the early 1970s. The papers at these meetings applied economic analysis to a wide range of issues.

As we began to think systematically about the determinants of the size or growth of government, we noted that many previous efforts had been based on specific institutional features or some form of myopia (or lack of information) by one or another party. Observations suggested that government spending and taxes had grown as a share of GNP in all developed economies. This seemed to rule out institutional explanations, at least those specific to a country. Since the phenomenon was widespread, it also seemed to cast doubt on explanations that rely on myopia.

Economic reasoning suggested an explanation based on individual choice by voters who maximize subject to constraints. This is the standard approach, which has been fruitful in economics, applied in a new area. A problem in this new application arises from the presence among the constraints of social decisions about how much to spend and to tax. The decisions about taxes and public spending are made by voters using the rules of maximizing behavior. The model we developed brought individual and collective choice together and showed that, in a modern democracy, tax rates depend on the voting rule, the distribution of income, and the income of the median, or decisive, voter. Later we tested a linear approximation of the model and found, to the surprise of at least one of us, that the model was not rejected. Subsequently a third member of our group was attracted by the often difficult, but always intriguing, problems that arose in trying to expand the model of taxation and redistribution to explain other features of fiscal policy such as tax progressivity and intergenerational redistribution.

Governments often stamp "secret" or "classified" on documents. The policy process moves forward on paper. Information is a valuable, often critical, input in decisions. We tried to understand how policymakers use information. We found that differences in the information available to voters and government officials help to explain some features of the policy process that had long seemed puzzling. Economists had often noted in particular contexts that policymakers could achieve desired outcomes efficiently by announcing their objective and, with sufficient private incentives, allowing markets to achieve the desired outcome. Interest rates or exchange rate objectives are examples. Yet this rarely happens. Policymakers often conceal their objectives for money growth, interest rates, or exchange rates. Even when they announce their objectives, they may not seek to achieve them. Knowing the behavior of policymakers, members of the public may be skeptical about the value of announcements.

Our observations on the use of information in the political process suggested that variability and uncertainty would be reduced if reliable information was released in a timely way. We explored features of the political-economic process that might explain, as an implication of rational behavior, why policymakers prefer ambiguity or secrecy. We found that a policymaker may benefit from uncertainty. Differences between the policies that maximize voters' welfare and actual policies can arise if there are costs of acquiring information or restrictions on the flow of information. The costs arising from these informational differences are a burden for the voter that is often difficult to avoid. We called these costs the "costs of democracy." The costs are gross costs; in practice, they would be measured relative to a system of unanimous consent, full information, or other idealization.

The two topics in political economy on which we have worked—taxation and redistribution; and information, credibility, and costs of democracy—are part of a much larger set. Economics and political science each offer different perspectives on these and other policy issues. Economics typically neglects the influence of voting and voters. Political science typically neglects the role of effort, incentives, and individual maximizing behavior.

Working on the chapters in this volume has been a rewarding intellectual odyssey in two respects. First, we have come to believe that the tools of economic analysis can be applied fruitfully to problems involving political decisions. Combining voting and majority rule with individual maximizing behavior seems a necessary path toward a more powerful theory of economic policy. Of course, we are not the first to make this discovery. Many have gone before. But doing the work has brought a sharper vision of the opportunities from this expansion of economics into adjacent areas of inquiry. Second, we have developed a better understanding of the powerful role of information in the policy process and the way in which possession of information creates opportunities and burdens.

Much of the work in this volume was done at the Graduate School of Industrial Administration, Carnegie Mellon University, between 1978 and 1989.

Several chapters have been published elsewhere, but Chapter 5 has not. It extends the model of taxes and redistribution to show when voters choose progressive income taxation. We have added a small section to Chapter 3, "Tests of a Rational Theory of the Size of Government," to include some additional evidence based on data for the United Kingdom.

While writing these chapters we incurred a number of intellectual debts that we can only partially discharge by acknowledgment. We owe a special debt to the late Karl Brunner, to whom the volume is dedicated. The participants of annual Interlaken seminars commented on several of the chapters in early stages and, through their questions and discussions, helped us to clarify many of the issues. Two of the chapters were presented as papers at the annual Carnegie Mellon Conference on Political Economy, where participants provided active and lively discussion of the issues. Colleagues at Carnegie Mellon and Tel-Aviv Universities made many contributions. We hope that we have recognized the contributions of our colleagues, and others, in the footnotes to the original papers. Lastly, we thank Alberta Ragan for preparing many drafts of the papers, and later the chapters, with skill, accuracy, and unflagging goodwill.

We are grateful to our colleagues and to the school for maintaining an environment conducive to research. We are grateful also to the Lynde and Harry Bradley Foundation, the Alex McKenna Foundation, the John M. Olin Foundation, and the Sarah Scaife Foundation for their generous support of our work.

Pittsburgh   A. H. M.
March 1991   A. C.
    S. F. R.

Acknowledgments

Chapter 1 is based on "Voting Rights and Redistribution: Implications for Liberal Democratic Governments" presented at the Conference on Liberal Democratic Societies, London, August 1989. It is reproduced with permission of the conference's sponsors, the Professors' World Peace Academy.

Chapter 2 is reprinted with permission from the *Journal of Political Economy*, *89*, pp. 914–27, copyright 1981 by the University of Chicago Press.

Chapters 3 and 4 are reprinted with permission of Kluwer Academic Publishers from *Public Choice*, *41*, pp. 403–18, and *47*, pp. 231–65, copyright 1983 and 1985, respectively, by Martinus Nijhoff Publishers.

Chapter 6 is reprinted with permission from *American Economic Review*, *79*, pp. 713–32, copyright 1989 by American Economic Association.

Chapter 7 is reprinted with permission from *Economic Inquiry*, *24*, pp. 367–88, copyright 1986 by Western Economic Association International.

Chapter 8 is reprinted with permission from *Econometrica*, *54*, pp. 1099–1128, copyright 1986 by The Econometric Society.

Chapter 9 is reprinted with permission from *Monetary Policy and Uncertainty*, M. J. M. Neumann (ed), pp. 39–67, copyright 1986 by Nomos Verlagsgesellschaft, Baden-Baden.

Contents

1
Introduction /3
ALLAN H. MELTZER

2
A Rational Theory of the Size of Government /23
ALLAN H. MELTZER AND SCOTT F. RICHARD

3
Tests of a Rational Theory of the Size of Government /36
ALLAN H. MELTZER AND SCOTT F. RICHARD

4
A Positive Theory of In-Kind Transfers and the Negative Income Tax /53
ALLAN H. MELTZER AND SCOTT F. RICHARD

5
A Political Theory of Progressive Income Taxation /76
ALEX CUKIERMAN AND ALLAN H. MELTZER

6
A Political Theory of Government Debt and Deficits in a Neo-Ricardian Framework /109
ALEX CUKIERMAN AND ALLAN H. MELTZER

7
A Positive Theory of Discretionary Policy, the Cost of Democratic Government, and the Benefits of a Constitution /135
ALEX CUKIERMAN AND ALLAN H. MELTZER

8
A Theory of Ambiguity, Credibility, and Inflation under Discretion and Asymmetric Information /158

ALEX CUKIERMAN AND ALLAN H. MELTZER

9
The Credibility of Monetary Announcements /191

ALEX CUKIERMAN AND ALLAN H. MELTZER

Index /215

POLITICAL ECONOMY

1

Introduction

ALLAN H. MELTZER

"Individuals have rights, and there are things no person or group may do to them (without violating their rights). So strong and far-reaching are these rights that they raise the question of what, if anything, the state or its officials may do." Nozick's 1968, defense of individual freedom opens with these words (p. ix). What do the words mean?

As a positive proposition about the limits to government's control or influence over property, income distribution, and the allocation of resources, Nozick's statement is rejected everywhere. In countries where people choose their political leaders directly or indirectly in competitive elections, what government "may do"—its command over resources through taxation, regulation, or other means—has expanded for at least a century. Redistribution across groups, both temporally and intertemporally, is the major reason for the growth of government budgets in developed market economies with near universal suffrage (see Nutter, 1978). In totalitarian systems, where competitive elections are restricted or prevented, as in China and until recently Eastern Europe and the Soviet Union, most property rights vanish, and the state is the dominant influence on income distribution or redistribution.

Nozick's proposition also fails as a relevant statement about what governments ought to do or not do about resource allocation in practice. The reason is that the proposition has no application to totalitarian societies, and it does not take account of voting. Neither Nozick's statement nor the text that follows it proposes a means by which voters can restrict the role of government to activities that do not violate some individual's rights to property and income. Evidence suggests that no major political group supports Nozick's proposals. In many countries liberal parties that once exhorted voters to protect property

---

I have benefited from many discussions with Karl Brunner, Scott Richard, and Alex Cukierman over many years. Dennis Epple, Robert Hahn, William Meckling, Steve Pejovich, and Keith Poole made helpful comments on an earlier draft. A previous version of this chapter was presented at the London Conference on Liberal Democratic Society, the Karl Brunner Seminar at Interlaken, and the Hayek Seminar.

rights and restrict the role of government have either declined to insignificance in this century or changed their programs.

The assignment of tasks to a collective body—the state—and the use of coercion has generated a large literature. A problem with much of this literature is that, like Nozick's book, it does not take account of voting. Near-universal suffrage provides voters an opportunity to redistribute property, income, and wealth. Once voting rights are granted, voters have the *right* to redistribute, subject to some limitations in law or custom. Often the limitations are reinterpreted or removed when there is sufficient public support for change.

This book takes a positive approach to the role of government in redistribution. In contrast, much previous literature on the role of the state is normative and, most recently, has been written from a constitutional or contractarian perspective. The following section reviews a small part of this literature to highlight differences in the approach taken here. Then I discuss the process of change and its relation to voting, coercion, and redistribution and summarize some earlier work on the relation of voting rights to property rights and redistribution. The discussion provides a frame of reference that has guided the work in the following chapters, although the latter are, of necessity, more restricted. A summary of the chapters brings out some of the main points of the more rigorous development that follows.

I. Constitutions, and Contracts

Political economy, as used here, is about the interrelation of individual and collective decision making. This involves the functioning of the market process, the political process, and the interaction between the two. Perceived failure of the market process, as in the depression of the 1930s, was followed by restrictions on markets; tax rates and redistribution increased; a search for alternatives to the market most often shifted decision making from the individual to some collective body. Perceived failure of socialist economies and sustained growth of the market economies, as in the past 40 years, has been followed by increased reliance on individual decisions and the market process. In democratic countries, shifts of this kind reflect the voters' consensus; they are based on the consent of the governed or, at least, of those who vote. A less orderly process occurs in nondemocratic countries where political outcomes are often less predictable and chosen by despots or tyrants. Physical force has been used to maintain tyrannies, and it is far from certain that force will not be used again in Eastern Europe, as in China, to prevent voters from making social decisions.

A key difference between democratic and nondemocratic processes governing change arises from the presence or absence of an accepted constitutional order. An operative constitution specifies how changes in rights and obligations may occur legally, which is to say in a way that has been agreed to or is acceptable to the relevant group. An operative constitution provides a means by which citizens can change the allocation of rights, including the right to vote and participate in the process of change.[1]

The theory of constitutional order has been treated in philosophy and in the social sciences, most recently in work on public choice or political economy. In his Nobel lecture, James Buchanan (1987) pointed out that, at the turn of the century, Knut Wicksell (1958) was one of the first to provide a foundation for choices involving individual and collective action. That foundation has three elements. *Methodological individualism* recognizes that choices are made by individuals acting according to their own preferences under constraints, including rules or social norms. *Economic man* posits that individuals make the choices purposefully to achieve their objectives. *Politics as exchange* treats the political process as a means of achieving private objectives through collective decisions. Wicksell retained the criterion of efficiency as a basis for judging outcomes of the political process, so he favored unanimity, or near-unanimity, in political decisions.

Two of Buchanan's contributions were to revive concern for these issues in economics and to shift the focus of discussion from policy actions or decisions to the rules under which the decisions are taken. Buchanan analyzed constitutions as rules that specify the process by which societies make or change the arrangements under which they are governed. If the rules are tightly drawn and enforced, they reduce uncertainty about whether property will be confiscated or wealth redistributed differently in the future than in the past and present. Rules provide for due process and restrict what governments can do. They determine who votes, how the rules for voting and for changing the constitution can be altered, and how rules may be enforced and interpreted.

A social order that avoids both the tyranny of a minority and mob rule restricts the actions of the government and the governed. What principles can, or should, govern how such restrictions are chosen? John Rawls (1971) proposed two principles that have received much attention. First, Rawls treats the constitution as a contract, voluntarily entered into, that is binding once it is accepted. Second, he assumed that agreement is reached behind a "veil of ignorance"; individuals choose the rules for redistribution before they know their income, wealth, or position in society.

Rawls and those who use the Rawlsian approach take a contractarian position. For them the constitution is the result of a voluntary agreement that is binding on the participants. The contractarian position obviously fails as a positive explanation of the development of states and constitutions. Nation-states did not evolve in this way, and constitutions are subject to change. As a normative statement, without foundation in a relevant, positive framework, the contractarian argument lacks a clear indication as to where it applies. If individuals have the right to vote, they can alter their prior decisions, for example by choosing higher taxes and more redistribution in one case or lower taxes and less redistribution in another. Such changes are the outcome of a

---

1. The emphasis on operative constitution is important. Countries like the Soviet Union have constitutions, but they are not binding on the rulers. Countries like Britain do not have written constitutions, but the constitution is operative; rulers are restricted.

constitutional process—for example majority rule—that specifies how changes can be made. A similar process, possibly requiring a super majority in place of a simple majority, permits voters to change the constitution. If, at some future date, voters change the voting rule to enfranchise persons who previously could not vote, both the voting rule and the extent of redistribution may change. The constitution is altered by a process that is part of the contract. The normative proposition that this change is unjust neglects prior agreement on the process by which the rules can be changed. And the process by which rules are changed can be changed also.

For Rawls, justice is a state or outcome achieved by a process that maintains the welfare of the participants when judged by a particular principle or rule for redistribution. Rawls' (1971) rule maximized the position of the poorest in society.[2] Others have modified his assumed social welfare function but have retained the veil of ignorance, the contractarian foundation, or both.[3]

Buchanan and Tullock (1962) use a concept of uncertainty similar to Rawls' veil of ignorance to develop the case for constitutional rules. Buchanan and Bush (1974, p. 153) are explicit that for them the "veil of ignorance" is a positive proposition about an individual's position when making binding choices of rules. However, unlike Rawls, they recognize not only that agreements may be broken once information about income, wealth, or status is revealed but also that individuals will take into account the prospect that some rules will be broken. Since many different rules for redistribution may be consistent with majority rule, the rule that survives must have some means of enforcement.

Brennan and Buchanan (1985) define a constitution as a set of rules governing procedural matters, specifying rights, and restricting the actions of governments. People establish governments "for the purpose of guaranteeing and protecting the rights agreed on in the contract" (p. 26). *All* members of the groups must accept the contract; the rules are made binding by unanimous consent (p. 27). A broad agreement of this kind on the constitution creates a public good.

To avoid some pitfalls of the contractarian position, Brennan and Buchanan treat the term *contract* as a metaphor for the process by which the constitution is accepted by the public. They explicitly reject the idea that the contract emerged or was universally accepted at a particular time. But they continue to use the Rawlsian veil of ignorance as applicable to rules. They argue that people choose rules unaware of the consequences that will follow from their choice but willing to accept those consequences.

Actual choice of rules is a continuing process, not a one-time event. There is no need to invoke the artificial assumption of a veil of ignorance to recognize our inability to predict the substantive outcomes that follow from any particular set of rules.

---

2. In 1974 Rawls offered a defense of this principle—the maximim principle—against maximizing average utility consistent with equality of opportunity.

3. See Musgrave (1989) and Buchanan (1987) as examples where parts of Rawls' framework are accepted explicitly.

The problems of choosing, sustaining, or enforcing rules for redistribution raise difficult and as yet unsolved problems of political economy. Notions of justice, equity, and fairness attached to some principles for redistribution should be seen as appeals for social consensus about decisions to tax and spend. Without some agreement on the outcome of the voting rule that is acceptable or enforceable and about what is or is not enforceable, it is difficult for heterogeneous individuals to reach a consensus.

In the standard economic welfare analyses of taxation and transfers, considerations of justice and fairness are never explicit. Optimal tax and spending decisions are derived by maximizing welfare subject to a budget constraint. The process of achieving consensus or majority decision and the institutions of government do not restrict decisions. Choices are made by a government of goodwill that seeks to achieve only efficient outcomes. Works by Becker (1983) and by Shepsle and Weingast (1984) are part of a growing literature that considers explicitly some of the deadweight losses arising from institutional practices such as the activities of pressure groups and logrolling in the Congress. Earlier, Niskanen (1971) used bureaucrats' desire for increased span of control as another mechanism generating excess burden.

By introducing relevant institutional features, models of this kind move away from the relatively barren theory of welfare economics. Logrolling, pressure groups, legislative staff, and bureaucrats are features of democratic government in many countries. Models with these features help to explain why there are pressures to expand the supply of public services and spending for redistribution in many countries. But growth in spending for redistribution has occurred in many countries operating under widely different governmental structures. Since these models lack voters, they do not explain why voters accept or choose redistribution.

Models that appeal to costs of information do not justify their neglect of voting. The reason is that often voters do not require information about the details of specific programs. Past experience can be informative. Upon announcement of programs for redistribution, public education, health, or child care, voters with relatively high income, or without children, knowing the past can be reasonably certain that they will bear a net cost. Voters with moderate or low incomes or large families know that their net benefit is likely to be positive. Upper income voters are usually represented by a party or parties that speak in favor of limited government, efficiency, growth, and incentives. Representatives of lower income voters advocate social justice and fairness, by which they often mean redistribution. Voters' agents, including elected representatives, understand how redistributive programs coerce some voters to incur costs in excess of benefits, in order to provide net benefits to others.

My purpose is not to deny that institutional structure can influence specific outcomes and encourage coercion and redistribution. As Buchanan has emphasized, the problem is to explain why voters sustain such outcomes when faced repeatedly with choices. A full explanation must include voter response and the choice of rules governing legislative or administrative processes. Ignorance, myopia, and lack of information are not sufficient to explain why rational voters in many countries have permitted redistribution and taxation to rise from decade to decade for a century.

Hayek has given extensive consideration to the principles of social organization and their relation to freedom, liberty, and progress in an uncertain world. From the *Constitution of Liberty* (1960) to *The Fatal Conceit* (1988), he analyzed these relations, emphasizing the dynamics of change and the role of chance. In Hayek's works, societies evolve in response to unforeseen and unpredictable changes. Each new circumstance gives rise to incentives and opportunities that cannot be foreseen in their entirety by anyone, so the consequences of rules or constitutions differ from those that are expected by their authors.

Economic and social development in this framework becomes a process of trial and error, groping and searching, accepting and discarding. In a Hayekian world, search, adaptation, and adjustment replace Rawls' static veil of ignorance. Progress consists of abandoning rules that produce outcomes that restrict the liberties enjoyed by members of the group. Hayek emphasized that "evolution cannot be just" (1988, p. 74) if justice is defined as a state or outcome as in Rawls. The environment in which decisions are taken changes; standards of judgment adapt or change. A static rule for income distribution, as in Rawls, will produce different levels of satisfaction and acceptance at different times.

Hayek's choice of liberty as a criterion introduces a normative element; in practice, liberty has not been the only principle guiding change. Alchian (1950/1977) proposed a broader principle for a particular set of institutions. He suggested that survival is the standard by which to judge the success and failure of business firms. Under conditions of uncertainty, distributions of returns from different strategies or decisions generally overlap. Individuals and firms make choices using available information, including knowledge of the outcomes of previous decisions made by others. This knowledge provides opportunities to adopt successful strategies or to attempt to improve on them. Some strategies will succeed by chance, so copying or continuing a previously successful strategy does not guarantee success.

With uncertainty, some firms' decisions will be profitable by chance, and others will produce losses by chance. Observers, using this information, will make errors—adopting decisions or arrangements that will not be profitable and rejecting arrangements that could succeed. In the long run, the effects of chance are minimized. Successful strategies persist and are profitable. Survival becomes the test of success.

Brunner and Meckling (1977) extended and applied Alchian's analysis to other institutions as part of what they called the REMM model. Resourceful, evaluating, maximizing man (REMM) makes decisions under uncertainty, using available information. He strives to improve—to maximize utility—given his preferences and the circumstances he faces. In this model, as in Alchian's, long-run survival of an institution is the measure of success. Success of an institution or arrangement such as the firm, the family, or a voting rule is, of course, always relative to alternative arrangements.

The emphasis on the long run is critical. In all social processes there are short-run costs of change and costs of acquiring information. Changes are difficult to interpret with precision, in part because it is difficult to control environmental or background conditions, but also because it is difficult to separate persistent from temporary changes in the environment. The social sciences and theories of stochastic processes are not yet able to specify the relevant conditions to be held constant or to identify the shape of the distribution of changes applicable to many choices. In the long run, these factors have less significance. Rules and institutions survive if voters or consumers prefer the outcomes they help to generate.

II. Coercion and Change

Hayek's (1960) definition of a liberal or free society is a society without coercion. Coercion occurs when "one man's actions are made to serve another man's will" (p. 133). Hayek treats a free society as an ideal that cannot be fully realized in practice. Everyone will not voluntarily agree to pay his share of the cost of collective actions that improve welfare. Where unanimity is not achieved, force, or threat of force, must be used to collect taxes and defend freedom or liberty. There is no operational principle that separates free riding from action's that "serve another's will." Friedman (1962, Chap. 2) would limit the role of government to maintaining law and order (to prevent coercion); enforcing contracts; defining, interpreting, and enforcing property rights; and providing a monetary framework. He recognizes the problems of monopoly, where technical considerations permit only a single producer, and neighborhood or third-party effects, but he draws no general conclusion about the role of government in these cases.

The recommendations of Friedman and Hayek, and many similar recommendations, are normative statements about the properties of a liberal society. In practice there is a conflict between democracy and the requirements for a liberal, noncoercive society. Democracy gives voters some rights to require all members of the community to pay for goods, services, and redistribution that some do not want and would willingly forego. Neither the decisions to spend nor the rules or constitution under which the decisions are made can be regarded as a contract freely entered into and to which all have consented. On the contrary, many would like to change the rules and the decisions but they are unable to do so.[4] The only alternatives open to them are other societies with different degrees of coercion, since there are no fully liberal societies in the sense of Hayek or Friedman.

The conflict arises because a liberal society requires near unanimity, and all democratic societies have some type of modified majority rule. Observation suggests that where governments do not allow the majority to rule, force must be used to maintain authoritarian control. Either voters have the authority to exercise control by choosing outcomes and public officials, or they may be

4. The changes can be toward more or less coercion and redistribution, as discussed more fully below. In general, people have migrated to societies with less coercion and fewer restrictions.

prevented from choosing by physical force or threat of force. An authoritarian government that is unwilling to use force gives way either to another authoritarian government or to some type of rule by the voters.

Recent history provides many examples. Stalin used force and the threat of force to maintain compliance with his decisions. In 1989, events in South Africa, China, and Eastern Europe showed that a perceived reduction in the threat of physical force leads to demands for a more democratic government. But China subsequently demonstrated also that the alternative to permitting democratic rule is a system maintained by physical force and the threat of force. The evidence, old and new, is so extensive and well known that further discussion seems unnecessary to establish that, if they are given the opportunity, people choose to make collective decisions by some type of majority rule with near universal suffrage and one vote per adult person. Other voting rules do not survive.

Majority rule is rarely if ever absolute, however. Restrictions are of various kinds. The legislative branch is often bicameral, with one chamber based on geographic location, personal standing, inheritance, or some other criterion that overrepresents particular groups. The members of the more popularly elected branch may win their seats in the legislature by a plurality rather than a majority, and the districts they represent may not be equal in population. The chief executive may be chosen by the legislature in parliamentary systems, by the electoral college in the United States, or by some other indirect method.[5] A judiciary, appointed for life, often has power to nullify laws and broaden or restrict their application. Changes in a written constitution, and passage of specific types of legislation and treaties with foreign governments, may require super majorities. These and other restrictions on rule by a simple majority are often intended to protect minority rights, including rights to property and income, or to reduce the influence of momentary passions or "mob rule."

Differences in the methods of choosing officials and in the organization of government may be important for understanding particular outcomes in particular countries. Observation suggests that majority rule produces income redistribution in all democratic countries; voters do not restrict the role of government to supplying public goods or eliminating market failures so as to equate private and social costs and benefits. Governments intervene with tariffs and restrictions on internal and external trade. Governments impose distortive taxes, subsidies, and regulations, and supply goods or services that can be produced privately. Education, health care, and pensions are examples of goods supplied by governments in all developed, democratic countries, but governments also collect garbage, treat sewage, and supply electricity in many countries. In each of these examples—and many others—part of the purpose may be to redistribute the costs and benefits of the good or service. Often greater efficiency would be achieved if redistribution were treated separately and private production were substituted for public production of the good or service.

---

5. Buchanan and Tullock (1962) argue that it is efficient to use different voting rules for choosing members of different branches.

Governments in many countries also set minimum or maximum prices for agriculture, labor, interest rates, and other goods or services. Again, the redistribution of wealth or income achieved by these programs could usually be achieved with smaller efficiency loss; there is an excess burden.[6] The provision of goods and services, particularly the public supply of private goods and services, contributes to the building up of bureaucracies. Costs of monitoring and obtaining information increase. Opportunities arise for public officials to mislead the public by issuing ambiguous statements or by concealing information. These, too, are costs of democratic government.

The evidence, of which these examples are a part, supports two propositions. First, although there are many differences in the detail of regulations and restrictions, there are many similarities across countries in the choice of goods and services that are publicly supplied or regulated. Second, voters redistribute income in ways that are costly. They do not choose least-cost redistribution. The inefficiencies are part of the cost of democracy. Later chapters analyze some of these costs formally.

III. Choosing and Changing the Rules

There are two possible interpretations of the outcomes of the political-economic process. Contractarians take the position that the rules are part of a contract to which all individuals consent. Since some vote against these outcomes, or against candidates who favor the outcomes, it is clear that consent to the outcomes is not unanimous. Further, since it is the rules under which they are governed that repeatedly produce the outcomes they dislike, voters must learn that remedy lies in changing the rules or constitution. Hence, all do not agree to the contract. To assume otherwise denies that people learn to associate outcomes with the rules under which the outcomes occur.

The economic model relies on unanimity or near unanimity, as Wicksell (1958) noted. The political economy model starts from the view that neither political decisions nor constitutions are based on unanimity or near-unanimity. All societies have some degree of coercion; the relevant set excludes non-coercion. In democracies, "consent of the governed" means rule by a (modified) majority (or super majority) that chooses and changes the rules or constitution. Those who oppose the rules may still decide to live under the rules in preference to available alternatives. In this limited sense they may be said to consent. Notwithstanding this weak form of consent, some will work to change the rules.

Once we admit costs of information and costs of organizing to protest, it becomes useful to distinguish between active support of the rules (or decisions) and passive acceptance. The former uses resources whereas the latter does not. A consensus in support of rules or decisions includes those who organize and actively support the decision or rule and those who take no action to oppose. In addition, there are those who dissent; this group actively opposes rules or decisions and works for change.

6. This view differs from that of Becker (1983).

The Vietnam War protest is a recent example. A relatively small number of dissenters worked to change the minds of many who initially did not dissent from the government's position. By focusing attention on their protests, they helped to change many passive supporters into active opponents and thus influenced the policy. The dissenters recognized that costs of distributing information can be reduced by marches, public rallies, draft card burning, and other activities that attract media attention. The change was not limited to the particular decision. Enough members of Congress accepted the general position to change the policy toward military intervention for at least a decade and to pass legislation like the War Powers Act limiting presidential discretion.

The same process has been used to change judicial decisions. Those who dissent from a Supreme Court decision rarely persuade a super majority to amend the constitution. There are exceptions such as womens' suffrage and prohibition. An alternative is to make an issue so prominent in national elections that a candidate may promise to appoint justices who are willing to reverse the decision. Abortion is a recent example of an issue where strong opinions and active organization by those who opposed the Supreme Court's *Roe v. Wade* decision had an effect through the political process. Gradually, the composition of the courts changed until the rule was changed. Political activity did not cease. A different group of active dissenters worked to change the rules.

These examples provide no support for the contractarian position. People do not unanimously consent to the rules for pornography, prohibition, abortion, school prayer, or war powers. Some favor a constitutional amendment to limit government spending or to require a balanced budget except in emergencies. On these and other issues there is no prospect of near-unanimity. Some live under laws and rules that they oppose. They are coerced to permit and finance activities of which they do not approve or are prevented from engaging in activities that they favor.[7]

At the state level in parts of the United States, and at the federal level in Switzerland, voters can use initiative or referendum to adopt or reject rules. They can change tax rates, spending, or regulation, so they can redistribute income. In the 1970s, voters in several states reduced tax rates and limited future increases. In 1988, a majority of the voters in California decided to reduce the price they pay for property and casualty insurance. They amended the constitution by initiative to reduce insurance rates by 20 percent and to remove insurance companies' exemption from the state's antitrust laws. The state Supreme Court upheld their right to do so, subject only to requirements of due process and "adequate" profit.

In each of these cases, and in many others, welfare economics has little to offer. The contractarian position fails and classical liberalism does not occur. In a system based on some form of majority rule, the ballot assigns rights, obligations, and wealth transfers in ways that depend on the applicable voting rule. Some individual rights and incomes are taken away; others are granted.

---

7. Decisions can require near-unanimity, but few do. An example is the requirement in the U.S. constitution that a state cannot be deprived of equal representation in the Senate without its consent.

Two processes are at work in democratic market economies. One is the market, where allocations are made by agreement and there is unequal distribution of wealth. The other is the political process, where allocations are made by majority rule and everyone has an equal vote. It is not surprising that the interaction of the two processes produces different outcomes and allocations from either process acting alone.

If the voters were unconstrained by incentives, or disincentives, and market responses and didn't care about the future, the political process would transfer wealth from the rich to the poor until incomes were equalized. Democratic market economies do not reach this outcome. The reason is that voters recognize the disincentives in any system of taxes and transfers that would equalize income.

If we assume that people use the same framework to make decisions in the polling place as they do in the marketplace, voters seek to maximize utility by allocating consumption over time. By raising taxes and increasing redistribution, a majority can increase their current consumption. Higher taxes, however, reduce investment and effort, so future income and consumption are reduced. If the majority vote to reduce taxes and transfers, aggregate consumption may increase while current consumption for the majority falls. The political problem is to find an equilibrium that the majority accepts. The idea, from welfare economics, of having the gainers compensate the losers would make the democratic process of redistribution futile. Instead, the majority chooses a tax rate that maintains equilibrium in the polling place and that redistributes income both currently and intertemporally.

The political-economic process produces a stable equilibrium outcome in many democratic countries, but there is no evidence suggesting convergence to a unique equilibrium tax rate common to all democratic countries. Actual tax rates, redistribution, and other features of the political-economic process differ over time and across countries. The political economy model implies that the differences should be related to differences in voting rules, distributions of ability and productivity, past experiences as they affect attitudes to work and leisure, and incentives or disincentives arising from regulation, details of tax and subsidy arrangements, and the like.

Some examples illustrate how differences may arise. Switzerland and Japan have had very stable political systems. The same party (or parties) has been elected to fill major offices for many years. Switzerland also gives voters the opportunity to vote directly in referenda. These features of Japanese and Swiss political arrangements are consistent with greater stability in the share of taxes paid and government spending for redistribution than is true in countries like Germany, Britain, or the United States, where political parties representing different majorities (or pluralities) alternate in office.

Larger differences become apparent if we compare Belgium, Canada, Lebanon, and Switzerland. Each has a relatively small population and has major religious and cultural differences. The political-economic outcomes in these countries range from highly stable to intermittent civil war. Farther removed are the so-called socialist countries where political power is concentrated narrowly; the public has little influence on the political process; political decisions are taken by a small group and are difficult to change by orderly political activity. Still, incentives or disincentives operate, and efficiency is often low. Once tyrannical restrictions began to relax, people demanded the right to vote in contested elections.[8]

The political economy model of taxes and redistribution suggests that, given the distribution of world income, an agreement permitting the United Nations to levy taxes by majority rule in the General Assembly would likely have effects broadly similar to an extension of the franchise. Redistribution from relatively wealthy to poorer countries would increase. Also, the model suggests that the stability of the political-economic equilibrium depends on the distribution of income. If available data are reliable, in Peru and Brazil the distribution of income is relatively skewed. There are comparatively many low-income recipients and comparatively few high-income earners, and comparatively large differences between mean and median incomes. Universal suffrage with majority-rule voting has proved to be unstable in these countries. With universal suffrage, the income of the median voter is farther below the mean than in countries with similar per capita income. Pressures for redistribution raise tax rates, reduce incentives, and thus reduce growth of output and investment. Those with high productivity leave or send their assets abroad to escape taxation or confiscation. Spending then exceeds revenues; the deficit is financed at the central bank, so inflation increases. An authoritarian government may take charge to restore stability and reduce spending and inflation.[9]

The World Bank (1988) has provided data on income distribution in developing countries. The data are almost certainly crude, and the observations are for different years, so conclusions drawn from these data must be tentative. I have selected countries with per capita incomes in 1986 in the middle income range, between U.S. $800 and $2400. The World Bank does not give median income; instead I have summed the shares for the two lowest quintiles, the lowest 40 percent of the income distribution. Table 1-1 shows these data in two sections. On the left are countries that have income distributions similar to those of democratic developed countries. On the basis of only the distributional criteria, these countries should be capable of maintaining stable voting democracies. On the right are countries that, by this same criterion, may be unable to do so. Also shown is the year the income distribution was observed. Comparable data are shown in the footnote to Table 1-1 for three democracies, two with relatively high income and one with low income.

Many factors other than income distribution affect the survival of democratic government. Yet countries with great diversity in language, religion, and culture manage to prosper and remain democratic, while countries with seemingly greater homogeneity on these dimensions fail to do so. In the left column of Table 1-1 we find countries of three types. There are stable

8. I do not claim that a parsimonious model predicts or explains all of these differences.
9. Currently there are no models predicting the policies of the authoritarian government. It may choose policies that increase incentives and encourage expansion, as in Brazil in the 1960s, or it may accede to some of the pressures for redistribution, as in Argentina at various times.

Table 1-1. Share of income of countries in lowest 40% of income distribution, according to World Bank data[a]

Year	Potentially Stable	Share of Income (%)	Year	Unstable	Share of Income (%)
1970	Argentina	14.1	1972	Brazil	7.0
1971	Costa Rica	12.0	1973	Panama	7.2
1976–77	El Salvador	15.5	1972	Peru	7.0
1976	Korea	16.9			
1985	Philippines	14.1			
1973–74	Portugal	15.2			
1975–76	Thailand	15.2			
1973	Turkey	13.5			

[a]For comparison: 1975–76, India, 16.2%; 1979, United Kingdom, 18.5%; 1980, United States, 17.2%.

democracies like Costa Rica; countries that made the transition from authoritarian to democratic government in recent years, such as Portugal, Thailand, and Turkey; and several countries that are now in the process of transition to democratic government—Argentina, Korea, Chile, and the Philippines. On the right are countries that have had difficulty maintaining democratic governments but in which competitive elections and majority rule have been used in the 1980s to choose political leaders and programs.

These examples highlight some implications of a political-economy framework. An economic model without a voting rule specifies social or political choices as a possibility frontier showing tradeoffs. To find a preferred position, a social welfare function must be added. In the political-economy model with a voting rule, this is not so. The relevant social choice is the decision made by the voters. Once the voting rule is specified, the model generates an equilibrium outcome or path for that voting rule.[10]

IV. Applications

Positive political economy differs from standard welfare economics in two respects. First, the positive approach does not introduce a social utility function or rely on a social planner to determine the amount or kind of redistribution. Once a voting rule is selected, the equilibrium combination of output and redistribution is determined by the voters given their personal positions in the income or wealth distribution and some structural features of the economy. Second, countries do not adopt optimal taxes, optimal regulation, optimal tariffs, compensation arrangements, or other implications of standard welfare

---

10. The voting rule can be complex in dynamic problems. The judiciary can overturn a law favored by a majority, but the public can influence the choice of judges with a lag that may require decades. Formal solutions to problems of this kind are difficult to achieve.

analysis. Government, acting in response to voters, engages in redistribution and coercion. Governments grant and reassign rights. Taxes are distortive, and redistribution may not be paid in the form most preferred by the recipients, so there are disincentives and excess burdens.[11]

The chapters that follow develop and apply parts of positive political economy outlined in previous sections of this chapter to a range of problems involving individual and collective decisions. The decisions we consider are mainly of two types. First are choices about taxation and redistribution. Individuals' decisions to work or take leisure, and to spend or save and leave bequests, are affected by collective decisions which they make as voters. We analyze the interrelation of these individual and collective decisions when both are endogenous, and we derive equilibrium positions that depend on such political and economic initial conditions as the distribution of income and the voting rule. Second are policy problems, where information or its lack is of paramount importance. Policymakers reveal less than full information about their objectives, actions, or changes in policy objectives. Their actions and statements are ambiguous and not fully credible to voters. Voters use information efficiently, but they cannot be certain that their information is correct or complete. The costs arising from ambiguity and credibility about policy actions is part of the cost of democratic government.

The issues we treat are a small part of the spectrum of issues in political economy. We do not address several of the topics mentioned in earlier sections of this chapter—regulation, direct intervention, and tariffs. Our work on taxation and redistribution analyzes a long-run democratic equilibrium where voters are fully informed about the consequences of their choices. Taxes and spending are driven by voters' demand.

Previous sections discuss passive and active participants—passionate minorities and silent majorities. We believe this is a relevant feature of the political-economic process, but it has not been incorporated in our models. Similarly, logrolling, the interests of administration and staff, and other institutional details are neglected.

Our aim has been to develop frameworks applicable to problems that arise in many countries. The increased size of government in the postwar years is the issue that first attracted our attention to political economy. The size of government increased under very different political and administrative arrangements. Taxes rose as a share of income or output in countries with parliamentary or presidential systems, single-member constituencies or proportional representation, unicameral or bicameral legislatures, full-time or part-time legislators, mainly direct or mainly indirect taxes, and many other features. That the same event occurred in different institutional environments suggests a common process at work, even if the details of this process vary with time, place,

---

11. Browning and Johnson (1984) estimated the cost of redistributing income. The net cost depends on disincentives and varies with the elasticity of labor supply with respect to tax rates. They estimated that the cost of redistribution is relatively large per dollar redistributed. For moderate elasticities, Browning and Johnson reported the losses to the net payers are more than $9 for every dollar received by the net beneficiaries.

and institutions. Similarly, the problem of monitoring, assessing, and using information about policies or actions arises in all political and economic systems.

Chapter 2 sets out the basic model of redistribution. Individuals work to produce consumption goods or take leisure. Political activity is redistributive. Taxes are proportional to income, but redistribution is a lump-sum grant. Individuals differ in ability and productivity. Their abilities and efforts in the marketplace determine their incomes and, by aggregation, society's income. Decisions in the polling place determine the size of the government budget for redistribution and the taxes that must be paid to balance the budget. A general equilibrium model links these decisions given the distribution of income and majority voting.

The model implies that each person's decision depends on the collective decision. The reason is that everyone pays taxes, but total tax collections are set by voting. Since there is no uncertainty and no costs of information, voters recognize this interaction when making choices. Higher taxes reduce effort and output. The "poor" recognize that taxing the "rich" lowers the output of the most productive members of the community. This knowledge limits redistribution to the optimum chosen by a majority. At the optimum, voters with incomes above the median prefer lower taxes and less redistribution. Those below the median would prefer higher taxes and more redistribution. The former may complain about the "burden of taxation," the "insufficiency of incentives," the "excessive amount of redistribution," the "demise of liberalism," and the "triumph of egalitarianism." The latter group has the opposite complaints; they argue for greater "equity," more concern for "social justice," and increased taxation of the rich. Both groups are coerced by the voting rule (and the requirement that the budget be balanced in real terms) to accept an outcome that they would prefer to change. Observations suggest that, despite repeated complaints of both types, the tax rate and the share of income spent on redistribution change slowly. The equilibrium appears to be stable.

As in well-known versions of Wagner's law, the model implies that tax rates and redistribution increase with income. Long-term changes in taxes and spending result from higher income but also from changes in the voting rule and shifts in the distribution of income. Chapter 3 presents tests of the influence of income and income distribution for the United States. A postscript to the chapter adds tests for the United Kingdom based on data from tax returns. We find some support for the model. Government spending for redistribution and the share of aggregate spending distributed as cash transfers rise and fall with the ratio of mean to median income and the level of median income.

Many economists, following Friedman (1962), have proposed replacing the current welfare system with a negative income tax that would make cash payments to welfare recipients. The proposal, if adopted, would remove burdensome costs and permit recipients of transfers to choose the goods and services they prefer, increasing their utility.

The negative income tax has not been adopted anywhere. Cash payments are typically limited to persons who do not work for reasons of age, infirmity, or unemployment. A large share of the transfer budget pays for in-kind transfers of goods and services such as health, food, housing, and education. Rules restrict recipients from trading these goods and services to achieve a preferred bundle. Although the rules cannot be fully enforced, they have an effect.

Chapter 4 extends the model of redistribution to include two consumption goods and leisure. The chapter shows that reliance on in-kind transfers and rejection of a negative income tax cease to be puzzling. The appeal of the negative income tax to the recipients of transfers is not sufficient to gain adoption unless a majority of voters subsist on transfers and do not work. Typically, voters who work, not welfare recipients, are on the margin; they choose the type of redistribution that benefits them. In-kind transfers are preferred by a majority of voters because they increase work, output, and tax payments by the recipients. Cash payments of equal value would have lower deadweight loss in consumption but would reduce recipients' incentive to work.

Recent legislation is consistent with the model. States and the federal government have adopted work requirements. Introduction of an earned income tax credit encourages work by welfare recipients, as the model implies.

To finance spending for redistribution and other activities, governments use taxes that introduce distortions. Progressive income taxes, though widely used, are difficult to reconcile with welfare analysis. Widespread reliance on progressive taxes is puzzling, as Blum and Kalven (1953) noted long age. Some recent work (Koester & Kormendi, 1989) deepens the puzzle by showing that increases in marginal tax rates, with average tax rates held constant, lower average per capita income.[12] Why are voters in all democratic countries willing to lower income?

Chapter 5 shows that this puzzle, too, can be resolved in a political-economy model. We replace the linear tax function of Chapter 2 with a quadratic tax function. The decisive voter chooses the marginal and average tax rates and the amount spent for redistribution. With universal suffrage and mean income sufficiently above median income, a majority favors redistribution and progressive income taxes even if the choice is socially costly. A majority is willing to pay a positive price to increase its own current consumption through redistribution.

The tax schedules used in Chapter 5 permit voters to choose regressive, proportional, or progressive taxes. We show that, for local changes, the individual with median ability is decisive. A majority prefers the schedule it chooses to any change in the neighborhood of the political-economic equilibrium. To show that this schedule is also a global optimum, restrictions must be put on admissible utility functions and the middle class must be of sufficient size relative to the voting population.

We find many conditions under which taxes will be progressive. Several of the conditions have intuitive appeal.[13] Progressivity is more likely the larger the

12. For developed countries, Koester and Kormendi estimated the effect of a 1-percent increase in the marginal tax rate, given the average tax rate, as a 0.75-percent reduction in per capita income.

13. At a seminar at the University of Bonn in 1988, Dieter Bös pointed out that the tax schedule in our chapter is similar to the schedule recently adopted by the West German government.

spread of the distribution of income or in the presence of a sufficient number of high-income individuals whose hours and efforts are relatively insensitive to taxes.

A common practice in all developed countries requires the current generation of workers to provide pensions for retired persons. In return, they receive pensions that are paid by a future generation. This rule, or social contract, is clearly redistributive. Wolff (1987) found that for the United States the direction of redistribution for old age assistance was from workers to all recipients, but low-wage workers benefited disproportionately.

Chapter 6 analyzes the political economy of intergenerational redistribution. Rational maximizing voters determine the size transfer and the method of financing by majority rule. The programs transfer income to the current generation by increasing current consumption, thereby crowding out part of the capital stock. As Wolff found, the benefits go to the current poor whose consumption increases disproportionately.

Barro (1974) used an intertemporal model to show that relative prices and wealth are unaffected by debt-financed budget deficits. In this analysis individuals reverse privately any effects of debt or deficits on the consumption and saving of current and future generations. This conclusion is puzzling. Congress, Presidents, and large parts of the public express concern about debt and deficits, issues that do not matter according to Barro's analysis.

The political-economy model removes the puzzle. To bring out the reasons for differences in conclusions, our model remains as close as possible to Barro's. Differences arise because we allow people to differ in ability, and therefore in income, and also in their initial (human) wealth. There is no uncertainty, but there is growth. The current poor know that their offspring will be wealthier than they are, so they do not wish to leave bequests. They would like to leave debts (negative bequests), but the law does not permit individuals to leave net debts to their heirs. This restriction limits the opportunity for individuals, acting alone, to redistribute from the future to the current generation by spending more than their income and any inheritance they receive. We call such people "bequest constrained." We show that the bequest constrained can relax the restriction by using government debt to finance intergenerational transfers. If the bequest constrained are part of a majority, they can vote for a deficit financed by selling debt. This permits them to increase their consumption. By doing so, they crowd out real capital and raise the interest rate.

The coalition voting for deficits and debt includes both rich and poor. Rentiers benefit from the higher real interest rates, so they favor deficits. The bequest constrained smooth family consumption intertemporally. They gain by removing the constraint and increasing their consumption. Their gain comes at the expense of the future consumption of their progeny.[14]

Much of the discussion of constitutions in politics and philosophy concerns rules and methods of setting rules. Much of the discussion in economics concerns the difference between policy rules and discretionary action. Chapter 7

---

14. Seiglie (1990) tests the intergenerational redistribution hypothesis and finds support.

takes a positive approach by asking how policymakers gain from discretion. What is the source of their gain, and what is the cost to the public?

We assume that the policymakers' objective is to get reelected. Their chances of success depend on social welfare during their term of office. In the presence of uncertainty and imperfect information about their ability and the state of the economy, policymakers get some benefit from favorable events over which they have no control, and they are harmed by unfavorable events during their term. Imprecise and incomplete information gives the policymakers opportunities to take actions that improve their chances of reelection. These discretionary actions may differ from the actions the public prefers. The actions impose costs on the public.

We call the costs "costs of democracy" because they are the result of actions taken to improve reelection prospects. If there were no elections, the costs would not arise.[15] The costs increase with the frequency of elections, and they disappear if the public has the same information as the government.

Chapter 7 shows that a rule or constitutional commitment to socially optimal policies would eliminate discretionary action and this cost of democracy, provided the public adopts the rule and the rule can be enforced. Where individuals differ in their policy views, we show that it is difficult to obtain agreement on a constitutional rule. Differences in views and the opportunity for the policymaker to benefit from discretion, therefore, limit the domain of rules. Many government actions remain discretionary.

Chapters 8 and 9 continue the analysis of information by considering the credibility of monetary actions with and without public announcements of monetary targets. Policymakers have state-contingent objectives. Their objectives are serially correlated, but the weight on a particular objective, such as economic stimulation or avoidance of inflation, changes over time. The public does not know when shifts in objectives occur. They monitor the actions of the central bank and use all available information, but their information is incomplete.

The central bank's control of money is imperfect. Changes in money occur because the central bank changes its objectives but also because of control errors. Consequently, the history of money growth and current changes in the growth rate do not immediately reveal whether there has been a change in policy objectives. Central bankers know their current objectives, but they do not know when they will change.

We define credibility as the difference between the policymaker's plans the the public's beliefs about those plans. Credibility is a stock. Policymakers can draw on the stock, by inflating more than anticipated. By doing so, they reduce credibility but are able to expand output temporarily by monetary means. Policymakers pay for the reduction in credibility when they wish to reduce inflation. They must rebuild the stock by convincing the public that the new policy will remain disinflationary.

Imperfect credibility is not the result of dynamic inconsistency, as in

---

15. Obviously, these are not net costs.

Kydland and Prescott (1977). The credibility problem arises from the superior information of policymakers, his shifting objectives, and imperfect control.

Chapter 8 shows that policy has an inflationary bias. The poorer the monetary control, the larger the inflationary bias. Also, the model of Chapter 8 implies that monetary uncertainty may be above its minimal feasible value. Available technology sets a lower limit to uncertainty, but as shown by the Federal Reserve (1981), the Federal Reserve does not achieve the minimum control error. Poor control helps the Federal Reserve to create surprises.

Chapter 9 incorporates announcements of targets for money growth. Announcements provide additional information to the public, but the reliability and credibility of the information varies. Announcements are a noisy signal of the policymaker's intentions. The public benefits from the announcements, however, since the announcements have some, perhaps minimal, value.

Chapter 9 compares three monetary regimes. The first has discretionary policy action. The second adds monetary announcements but retains discretionary action. The third is a rule for constant money growth.

We show that constant money growth reduces monetary uncertainty by neutralizing the effect of changes in political priorities on money growth. The monetary rule provides the public with the most accurate information. We show, however, that if announcements are fully accurate and equal to the policymaker's planned rate of money growth, the uncertainty of discretionary policy with announcements falls to the level achieved under the monetary rule.

References

Alchian, A. A. (1950/1977). "Uncertainty, Evolution and Economic Theory." *Journal of Political Economy*, 58 (June). Reprinted in *Economic Forces at Work*. Indianapolis: Liberty Press, pp. 15–35.

Barro, R. (1974). "Are Government Bonds Net Wealth?" *Journal of Political Economy*, 82 (Nov./Dec.), pp. 1095–1117.

Becker, G. (1983). "A Theory of Competition among Pressure Groups for Political Influence." *Quarterly Journal of Economics*, 98 (Aug.), pp. 371–400.

Blum, W. and Kalven, H. Jr. (1953). *The Uneasy Case for Progressive Taxation*. Chicago: University of Chicago Press.

Board of Governors (1981). *New Monetary Control Procedures*, Vols. I and II. Washington, D.C.: Federal Reserve System.

Brennan, G. and Buchanan, J. M. (1985). *The Reason of Rules: Constitutional Political Economy*. Cambridge: Cambridge University Press.

Browning, E. K. and Johnson, W. R. (1984). "The Trade-Off between Equality and Efficiency." *Journal of Political Economy*, 92 (Apr.), pp. 175–203.

Brunner, K. and Meckling, W. (1977). "The Perception of Man and the Conception of Government." *Journal of Money, Credit and Banking*, 9 (Feb.), pp. 60–85.

Buchanan, J. M. (1987). "The Constitution of Economic Policy." *Science*, 236 (June 12), pp. 1433–36.

Buchanan, J. M. and Bush, W. C. (1974). "Political Constraints on Contractual Redistribution." *American Economic Review, Papers and Proceedings*, 64 (May), pp. 153–57.

Buchanan, J. M. and Tullock, G. (1962). *The Calculus of Consent*. Ann Arbor: University of Michigan Press.
Federal Reserve (1981). *New Monetary Control Procedures*. Vol. I. Washington, D.C.: Board of Governors.
Friedman, M. (1962). *Capitalism and Freedom*. Chicago: University of Chicago Press.
Hayek, F. (1960). *The Constitution of Liberty*. Chicago: University of Chicago Press.
Hayek, F. (1988). *The Fatal Conceit*. W. W. Bartley (ed.). Chicago: University of Chicago Press.
Koester, R. and Kormendi, R. (1989). "Taxation, Aggregate Activity and Economic Growth: Cross-Section Evidence on Some Supply-Side Hypotheses." *Economic Inquiry*, 27 (July), pp. 367–86.
Kydland, F. E. and Prescott, E. C. (1977). "Rules Rather than Discretion: The Inconsistency of Optimal Plans." *Journal of Political Economy*, 85 (June), pp. 473–92.
Musgrave, R. A. (1989). "The Three Branches Revisited." *Atlantic Economic Journal*, 17 (Mar.), pp. 1–7.
Niskanen, W. A. (1971). *Bureaucracy and Representative Government*. Chicago: Aldine.
Nozick, R. (1968). *Anarchy, State and Utopia*. New York: Basic Books.
Nutter, G. W. (1978). *Growth of Government in the West*. Washington, D.C.: American Enterprise Institute.
Poole, K. and Rosenthal, H. (1989). "Color Animation of Dynamic Congressional Voting Models." Pittsburgh: Carnegie Mellon University, Graduate School of Industrial Administration.
Rawls, J. (1971). *A Theory of Justice*. Cambridge, Mass.: Harvard University Press.
Rawls, J. (1974). "Some Reasons for the Maximin Criterion." *American Economic Review, Papers and Proceedings*, 64 (May), pp. 141–46.
Seiglie, C. (1990). "Deficits, Defense and Income Redistribution." Xerox, Economics Dept., Rutgers University, New Brunswick, N.J.
Shepsle, K. and Weingast, B. (1984). "Political Solutions to Market Problems." *American Political Science Review*, 78, pp. 417–34.
Wicksell, K. (1958), *Finanz Theoretische Untersuchungen*. Translated in *Classics in the Theory of Public Finance*, R. A. Musgrave and A. T. Peacock (eds.). London: Macmillan, pp. 72–118.
Wolff, N. (1987). *Income Distribution and the Social Security Program*. Ann Arbor: University of Michigan Research Press.
World Bank (1988). *World Development Report*. Washington, D.C.: World Bank.

2

A Rational Theory of the Size of Government

ALLAN H. MELTZER AND SCOTT F. RICHARD

In a general equilibrium model of a labor economy, the size of government, measured by the share of income redistributed, is determined by majority rule. Voters rationally anticipate the disincentive effects of taxation on the labor-leisure choices of their fellow citizens and take the effect into account when voting. The share of earned income redistributed depends on the voting rule and on the distribution of productivity in the economy. Under majority rule, the equilibrium tax share balances the budget and pays for the voters' choices. The principal reasons for increased size of government implied by the model are (1) extensions of the franchise, which change the position of the decisive voter in the income distribution, and (2) changes in relative productivity. An increase in mean income relative to the income of the decisive voter increases the size of government.

I. Introduction

The share of income allocated by government differs from country to country, but the share has increased in all countries of Western Europe and North America during the past 25 years (Nutter, 1978). In the United States, in Britain, and perhaps elsewhere, the rise in tax payments relative to income has persisted for more than a century (Meltzer & Richard, 1978; Peacock & Wiseman, 1961). There is, as yet, no generally accepted explanation of the increase and no single accepted measure of the size of government.

We are indebted to Karl Brunner, Dennis Epple, Peter Ordeshook, and Tom Romer for many helpful discussions and to the participants in the Carnegie-Mellon Public Economics Workshop, an anonymous referee, the editor, and the Interlaken Seminar for constructive comments on an earlier version.

In this chapter, the budget is balanced.[1] We use the share of income redistributed by government, in cash and in services, as our measure of the relative size of government and we develop a theory in which the government's share is set by the rational choices of utility-maximizing individuals who are fully informed about the state of the economy and the consequences of taxation and income redistribution.[2]

The issues we address have a long intellectual history. Wicksell (1958) joined the theory of taxation to the theory of individual choice. His conclusion—that individual maximization requires government spending and taxes to be set by unanimous consent—reflects the absence of a mechanism for grouping individual choices to reach a collective decision. Following Downs (1957), economists turned their attention to the determination of an equilibrium choice of public goods, redistribution, and other outcomes under voting rules that do not require unanimity.

Several recent surveys of the voluminous literature on the size or growth of government are now available (see Aranson & Ordeshook, 1980; Brunner, 1978; Larkey, Stolp, & Winer, 1980; Peacock, 1979).[3] Many of the hypotheses advanced in this literature either emphasize the incentives for bureaucrats, politicians, and interest groups to increase their incomes and power by increasing spending and the control of resources or rely on specific institutional details of the budget, taxing, and legislative processes. Although such studies contribute to an understanding of the processes by which particular programs are chosen, they often neglect general equilibrium aspects. Of particular importance is the frequent failure to close many of the models by balancing the budget in real terms and considering the effect on voters of the taxes that pay for spending and redistribution (see, e.g., Hayek, 1979; Niskanen, 1971; Olson, 1965). An empirical study by Cameron (1978) suggests that decisions about the size of the budget are not the result of "fiscal illusion," so the neglect of budget balance cannot be dismissed readily.

We differ from much of the recent literature in three main ways. First, voters do not suffer from fiscal illusion and are not myopic. They know that the government must extract resources to pay for redistribution. Second, we concentrate on the demand for redistribution and neglect any "public goods" provided by government (see also Peltzman, 1979). Third, we return to the earlier tradition of de Tocqueville ([1835] 1965), who associated the size of government, measured by taxes and spending, with two factors: the spread of the franchise and the distribution of wealth (property).[4]

---

1. All variables are real. There is no inflation. Budget balance means that redistribution uses real resources. Public goods are neglected.

2. Ideally the size of government would be measured by the net burden imposed (or removed) by government programs.

3. Larkey et al. (1980) include a survey of previous surveys. Mueller (1976) summarizes contributions by Downs (1957), Musgrave (1959), Olson (1965), Niskanen (1971), Buchanan and Tullock (1972), Riker and Ordeshook (1973), and others to such related topics as the determination of equilibrium collective decisions and the effects of government policies on the distribution of income.

Our hypothesis implies that the size of government depends on the relation of mean income to the income of the decisive voter. With universal suffrage and majority rule, the median voter is the decisive voter, as shown by Roberts (1977) in an extension of the well-known works by Hotelling (1929) and Downs (1957). Studies of the distribution of income show that the distribution is skewed to the right, so the mean income lies above the median income. Any voting rule that concentrates votes below the mean provides an incentive for redistribution of income financed by (net) taxes on incomes that are (relatively) high. Extensions of the franchise to include more voters below mean income increase votes for redistribution, and thus increase this measure of the size of government.

The problem with this version of the de Tocqueville hypothesis is that it explains too much. Nothing limits the amount of redistribution or prevents decisive voters from equalizing incomes or, at a minimum, eliminating any difference between their disposable income and the disposable income of those who earn higher incomes. Incentives have been ignored. Higher taxes and redistribution reduce the incentive to work and thereby lower earned income. Once we take account of incentives, there is a limit to the size of government. To bring together the effect of incentives, the desire for redistribution, and the absence of fiscal illusion or myopia, we develop a general equilibrium model.

Section II sets out a static model. Individuals who differ in productivity, and therefore in earned income, choose their preferred combination of consumption and leisure. Not all individuals work, but those who do pay a portion of their income in taxes. The choice between labor and leisure, and the amount of earned income and taxes, depend on the tax rate and on the size of transfer payments.

The tax rate and the amount of income redistributed depend on the voting rule and the distribution of income. Section III shows how income redistribution, taxes, and the size of the government budget change with the voting rule and the distribution of productivity. A conclusion summarizes the findings and main implications.

II. The Economic Environment

The economy we consider has relatively standard features. There are a large number of individuals. Each treats prices, wages, and tax rates as givens, determined respectively in the markets for goods and labor and by the political process. Differences in the choice of labor, leisure, and consumption and differences in wages arise solely because of differences in endowments, which reflect differences in productivity. In this section we extend this standard model to capture the salient features of the process by which individuals choose to

---

4. We are indebted to Larkey et al. (1980) for pointing out the similarity between de Tocqueville and the conclusion we reached in an earlier version and in Meltzer and Richard (1978). De Tocqueville's distribution of property finds an echo in the concerns about "mob rule" by the writers of the Constitution.

work or subsist on welfare payments, and we show the conditions under which these choices are uniquely determined by the tax rate.

The utility function is assumed to be a strictly concave function, $u(c, l)$ for consumption c and leisure l. Consumption and leisure are normal goods, and the marginal utility of consumption or leisure is infinite when the level of consumption or leisure is zero, respectively. There is no capital and no uncertainty.

The individual's endowment consists of ability to produce, or productivity, and a unit of time that he allocates to labor, n, or to leisure, $l = 1 - n$. Individual incomes reflect the differences in individual productivity and the use of a common, constant-returns-to-scale technology to produce consumption goods. An individual with productivity x earns pretax income y:

$$y(x) = xn(x) \tag{1}$$

Income is measured in units of consumption.

Tax revenues finance lump-sum redistribution of r units of consumption per capita. Individual productivity cannot be observed directly, so taxes are levied against earned income. The tax rate, t, is a constant fraction of earned income but a declining fraction of disposable income. The fraction of income paid in taxes net of transfers, however, rises with income.[5] There is no saving; consumption equals disposable income as shown in (2):

$$c(x) = (1 - t)nx + r \qquad c \geq 0 \tag{2}$$

If there are individuals without any ability to produce, $x = 0$, their consumption is $r \geq 0$.

Each individual is a price taker in the labor market, takes t and r as givens, and chooses n to maximize utility. The maximization problem is

$$\max_{n \in [0, 1]} u(c, l) = \max_{n \in [0, 1]} u[r + nx(1 - t), 1 - n] \tag{3}$$

The first-order condition

$$0 = \frac{\partial u}{\partial n} = u_c[r + nx(1-t), \ 1-n]x(1-t) - u_l[r + nx(1-t), \ 1-n] \tag{4}$$

---

5. Reliance on a linear tax follows a well-established tradition. Romer (1975) analyzed problems of unimodality using a linear tax and predetermined government spending. Roberts (1977), using a linear tax and a predetermined budget, showed that the median voter dominates the solution if incomes are ordered by productivity. Linear tax functions are used also when the social welfare function is used to determine the optimal tax (see Sheshinski, 1972). The degree to which actual taxes differ from linear taxes has generated a large literature. Pechman and Okner (1974) found that the tax rate is approximately constant. King (1980) wrote that most redistribution in the United States and the United Kingdom comes from the transfer system, not from the tax system. Browning and Johnson (1979) show that conclusions about proportionality of the tax rate depend heavily on assumptions used to allocate the burden of indirect business taxes.

determines the optimal labor choice, $n[r, x(1-t)]$, for those who choose to work. The choice depends only on the size of the welfare payment, r, and the after-tax wage, $x(1-t)$.[6]

Some people subsist on welfare payments. From (4) we know that $n = 0$ is the optimal choice at productivity level

$$x_0 = \frac{u_l(r, 1)}{u_c(r, 1)(1-t)} \tag{5}$$

Individuals with productivity below x_0 subsist on welfare payments and choose not to work; $n = 0$ for $x \leq x_0$.

Increases in redistribution increase consumption. For those who subsist on welfare, $c = r$, so $\partial c/\partial r = 1$. Those who work must consider not only the direct effect on consumption but also the effect of redistribution on their labor-leisure choice. The assumption that consumption is a normal good means that $\partial c/\partial r > 0$. Differentiating (4) and using the second-order condition, $D < 0$ in footnote 6, restricts u_{cl}:

$$\frac{\partial c}{\partial r} = \frac{u_{cl}x(1-t) - u_{ll}}{-D} > 0 \tag{6}$$

Consumption increases with r for both workers and nonworkers, provided consumption is a normal good.

The positive response of c to r takes one step toward establishing conditions under which we find a unique value of r that determines the amount of earned income and amount of redistribution for each tax rate. The next step is to show that normality of consumption is sufficient to establish that earned income (income before taxes) increases with productivity.

Pretax income is

$$y(r, t, x) = xn[r, x(1-t)] \tag{7}$$

People who do not work, $x \leq x_0$, have $y = 0$ and $\partial y/\partial x = 0$. For all others,

$$\frac{\partial y}{\partial x} = n + x\frac{\partial n}{\partial x} \tag{8}$$

The first-order condition, equation (4), yields

$$\frac{\partial n}{\partial x} = \frac{u_c(1-t) + u_{cc}nx(1-t)^2 - u_{cl}n(1-t)}{-D} \tag{9}$$

---

6. By assumption, u is strictly concave, so the second-order condition is negative and (4) defines a maximum. The second-order condition is

$$\partial^2 u/\partial n^2 = D = u_{cc}x^2(1-t)^2 - 2u_{cl}x(1-t) + u_{ll} < 0.$$

The sign of $\partial n/\partial x$ is indeterminate; as productivity increases, the supply of labor can be backward bending. Pretax income, $y = nx$, does not decline, however, even if n falls. Substituting (9) into (8) and rearranging terms shows that the bracketed term in (10) is the numerator of $\partial c/\partial r$ in (6). Hence, $\partial y/\partial x$ is positive for all $x > x_0$, provided that consumption is a normal good:

$$\frac{\partial y}{\partial x} = \frac{u_c(1-t)x + n[u_{cl}x(1-t) - u_{ll}]}{-D} > 0 \tag{10}$$

The final step in establishing that there is a unique equilibrium solution for any tax rate uses our assumption that leisure is a normal good. The government budget is balanced and all government spending is for redistribution of income. If per capita income is $\bar{y}$, then

$$t\bar{y} = r \tag{11}$$

Let $F(\cdot)$ denote the distribution function for individual productivity, so that $F(x)$ is the fraction of the population with productivity less than x. Per capita income is obtained by integrating:

$$\bar{y} = \int_{x_0}^{\infty} xn[r, (1-t)x]dF(x) \tag{12}$$

Equation (12) shows that per capita income, and therefore total earned income, is determined once we know x_0, t, and r. From (5) we know that x_0 depends only on t and r, and from (11) we know that, for any tax rate, there is at least one value of r that balances the budget.[7] If leisure is a normal good, the value of r that satisfies (11) for each t is unique.[8]

Once r or t is chosen, the other is determined. The individual's choices of consumption and the distribution of his or her time between labor and leisure are determined also. The choice of r or t uniquely determines each individual's welfare and sets the size of government.

III. The Size Of Government

The political process determines the share of national income taxed and redistributed. The many ways to make this choice range from dictatorship to

---

7. The left side of (11) is nonnegative and is a continuous function of r that is bounded by $t\bar{x}$, where $\bar{x}$ is the average of x.

8. The normality of leisure means that $\partial l/\partial r > 0$ and therefore $\partial n/\partial r = -\partial l/\partial r < 0$. Since

$$\frac{\partial \bar{y}}{\partial r} = \int_{x_0}^{\infty} x\left(\frac{\partial n}{\partial r}\right)dF(x) < 0$$

the left side of (11) is a strictly decreasing, continuous function of r. The right side of (11) strictly increases with r. This implies that there is a unique value of r that satisfies (11).

unanimous consent, and each produces a different outcome. We call each political process that determines the tax rate a *voting rule*.

In this section we consider any voting rule that allows a decisive individual to choose the tax rate. Two examples are dictatorship and universal suffrage with majority rule. A dictator is concerned about the effect of his decisions on the population's decisions to work and consume, but he alone makes the decision about the tax rate. Under majority rule, the voter with median income is decisive, as we show below. We then show that changes in the voting rules and changes in productivity change the tax rate and the size of government.

Decisive voters choose the tax rates that maximize their utility. In making this choice, the voter is aware that the choice affects everyone's decision to work and consume. Increases in the tax rate have two effects. Each dollar of earned income raises more revenue, but earned income declines; everyone chooses more leisure, and more people choose to subsist on redistribution. "High" and "low" tax rates have opposite effects on the choice of labor or leisure and therefore on earned income.

Formally, individuals are constrained to find a tax rate that balances the government budget, equation (11), and maximizes utility subject to personal budget constraints, equation (3). The first-order condition for the decisive voter is solved to find a preferred tax rate:

$$\bar{y} + t\frac{d\bar{y}}{dt} - y_d = 0 \tag{13}$$

where y_d is the income of the decisive voter.

Roberts (1977) showed that if the ordering of individual incomes is independent of the choice of r and t, individual choice of the tax rate is inversely ordered by income. This implies that with universal suffrage the voter with median income is decisive, and the higher his income, the lower the preferred tax rate. By making the additional assumption that consumption is a normal good, we have shown that incomes are ordered by productivity for all r and t. Combining Roberts's lemma 1 (1977, p. 334) with our results, we can order the choice of tax rate by the productivity of the decisive voter.[9] The higher an individual's productivity, the lower the preferred tax rate.

Figure 2-1 illustrates the proposition and shows the effect on the tax rate of changing the voting rule. The negatively sloped line is the relation between individual productivity, x, and the individual's preferred tax rate. This line need not be linear.

The maximum tax rate, $t_{\max}$, is chosen if the decisive voter does not work. An example is $x = x_{d1}$. In this case, $x \leqslant x_0$; the decisive voter consumes only r, so

---

9. The formal statement of the result is: Consider any two pairs (r_1, t_1) and (r_2, t_2). If $t_2 > t_1$, then for all $x:, x$ is indifferent between (r_1, t_1) and (r_2, t_2) implies that x' weakly prefers (r_2, t_2) to (r_1, t_1) for all $x' < x$ and x'' weakly prefers (r_1, t_1) to (r_2, t_2) for all $x'' > x$; x strictly prefers (r_1, t_1) to (r_2, t_2) implies that x'' strictly prefers (r_1, t_1) to (r_2, t_2) for all $x'' > x$; x strictly prefers (r_2, t_2) to (r_1, t_1) implies that x' strictly prefers (r_2, t_2) to (r_1, t_1) for all $x' < x$. Note that this result does not require unimodality of voter preferences for tax rates.

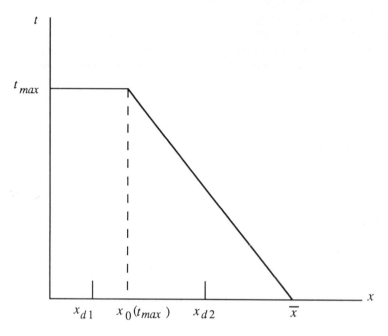

Figure 2-1.

the voter chooses the tax rate (t_{max}) that maximizes r. Any higher tax rate reduces aggregate earned income, tax collections, and the amount available for redistribution. From equation (5) we see that the maximum tax rate must be less than $t = 1$.

As productivity rises from x_0 to $\bar{x}$, the tax rate declines from t_{max} to 0. At $x_d = \bar{x}$, the decisive voter is endowed with average productivity and cannot gain from lump-sum redistribution, so he votes for no redistribution by choosing $t = 0$.[10] From equation (5) and $u_c(0, \cdot) = \infty$, we see that everyone works when $r = 0$. If the decisive voter's productivity exceeds $\bar{x}$, t and r remain at zero and aggregate earned income remains at society's maximum.

Changes in the voting rule that spread the franchise up or down the productivity distribution change the decisive voter and raise or lower the tax rate. Our hypothesis implies that changing the position of the decisive voter in the distribution of productivity changes the size of government provided $x_0 < x_d < \bar{x}$. Major changes in x_d have occurred in two ways. Wealth and income requirements for voting were reduced or eliminated, gradually broadening the franchise and lowering the income of the decisive voter. Social security retirement systems grew in most countries after the franchise was extended. By increasing the number of retired persons, social security systems increase the number of voters who favor increased redistribution financed by taxes on wages.

---

10. We have omitted public goods. In an earlier version we showed that under carefully specified conditions, public goods can be included without changing the result for redistribution.

Some of the retired who favor redistribution also favor low taxes on capital, property, and the income from capital.

The size of government changes also if there are changes in relative income, as shown by equation (13), or relative productivity. Conclusions about the precise effect of changes of this kind are difficult to draw. We cannot observe productivity directly and can only infer changes in the distribution of productivity, $F(\cdot)$, by observing changes in relative income. The literature makes clear that these effects are disputed (see King, 1980; Sahota, 1978; and others). Further, we cannot deduce the effect of changes in productivity on t directly from equation (13). The reason is that $\bar{y}$ depends on t, so finding the effect of changes in relative productivity requires the solution to a nonlinear equation in t. Instead, we rewrite (13) in a form that involves the (partial) elasticities of per capita income ($\bar{y}$) with respect to redistribution (r) and the wage rate ($x[1-t]$).

Let $\tau = 1 - t$ be the fraction of earned income retained. From (12), $\bar{y}$ depends on r and τ only. The total derivative is

$$\frac{d\bar{y}}{dt} = \frac{\bar{y}_r \bar{y} - \bar{y}_\tau}{1 - t\bar{y}_r} \tag{14}$$

where $\bar{y}_\tau$ and $\bar{y}_\tau$ are the two partial derivatives. Substituting (14) into (13), we solve for t:

$$t = \frac{m - 1 + \eta(\bar{y}, r)}{m - 1 + \eta(\bar{y}, r) + m\eta(\bar{y}, \tau)} \tag{15}$$

where m is the ratio of mean income to the income of the decisive voter, $\bar{y}/y_d$, and the η's are partial elasticities. Using the common economic assumption that the elasticities are constant, the tax rate rises as mean income rises relative to the income of the decisive voter, and taxes fall as m falls:

$$\frac{dt}{dm} = \frac{\eta(\bar{y}, \tau)[1 - \eta(\bar{y}, r)]}{[m - 1 + \eta(\bar{y}, r) + m\eta(\bar{y}, \tau)]^2} > 0 \tag{16}$$

Relaxing the assumption of constant elasticities weakens the conclusion, but we expect the sign of (16) to remain positive provided the change in the elasticities is small.

One of the oldest and most frequently tested explanations of the growth of government is known as *Wagner's law*. This law has been interpreted in two ways. The traditional interpretation is that government is a luxury good so that there is a positive relation between the relative size of government and the level of real income. Alt (1980) has questioned this interpretation of Wagner's idea. Alt (p. 4) notes that Wagner argued that there is "a proportion between public expenditure and national income which may not be permanently overstepped." This suggests an equilibrium relative size of government rather than an ever-growing government sector.

The traditional statement of Wagner's law—that government grows more rapidly than income—has been tested many times, but with mixed results. Peacock and Wiseman (1961), Cameron (1978), and Larkey et al. (1980) discuss these tests. Our hypothesis suggests that the results are ambiguous because Wagner's law is incomplete. The effect of absolute income on the size of government is conditional on relative income. Average or absolute income affects the elasticities in equation (15), and the relative income effect is given by m.

To make our hypothesis testable, we must identify the decisive voter. The applicable voting rule in the United States is universal franchise and majority rule. Under this rule, the voter with median income is decisive in single-issue elections, as we argued above. Hence the median voter is decisive in elections to choose the tax rate, so m is the ratio of mean to median income.[11]

IV. Conclusion

Government spending and taxes have grown relative to output in most countries with elected governments for the past 30 years or longer. Increased relative size of government appears to be independent of budget and tax systems, federal or national governments, the size of the bureaucracy, and other frequently mentioned institutional arrangements, although the relative rates of change in different countries may depend on these arrangements.

Our explanation of the size of government emphasizes voter demand for redistribution. Using a parsimonious, general equilibrium model in which the only government activities are redistribution and taxation, the real budget is balanced, and voters are fully informed, we show that the size of government is determined by the welfare-maximizing choice of a decisive individual. The decisive individual may be a dictator, absolute monarch, or marginal member of a junta.

With majority rule the voter with median income among the enfranchised citizens is decisive. Voters with income below the income of the decisive voter choose candidates who favor higher taxes and more redistribution; voters with income above the decisive voter desire lower taxes and less redistribution. The decisive voter chooses the tax share. When the mean income rises relative to the income of the decisive voter, taxes rise, and vice versa. The spread of the franchise in the nineteenth and twentieth centuries increased the number of voters with relatively low income. The position of the decisive voter shifted down the distribution of income, so tax rates rose. In recent years, the

---

11. The many tests of the median-voter hypothesis using regression analyses are inconclusive. One reason is that many of the tests do not discriminate between the median and any other fractile of the income distribution (see Romer & Rosenthal, 1979). Cooter and Helpman (1974) use income before and after taxes net of transfers to estimate the shape of the social welfare function implicit in U.S. data. They conclude that "the assumption that ability is distributed as wages per hour ··· — perhaps the best assumption on distribution of ability—vindicates the median voter rule."

proportion of voters receiving social security has increased, raising the number of voters favoring taxes on wage and salary income to finance redistribution. A rational social security recipient with large property income supports taxes on labor income to finance redistribution but opposes taxes on income from property. In our analysis, there is neither capital nor taxes on property, so the increase in social security recipients has an effect similar to an extension of the franchise.

Our assumption that voters are fully informed about the size of government differs from much recent literature. There, taxpayers are portrayed as the prey sought by many predators who conspire to raise taxes relative to income by diffusing costs and concentrating benefits, or in other ways (Buchanan & Tullock, 1962; Hayek, 1979; Niskanen, 1971; Olson, 1965). We acknowledge that voters are ill-informed about the costs of particular projects when, as is often the case, it is rational to avoid learning details. Knowledge of detail is not required to learn that the size of government has increased and that taxes have increased relative to output or income. Long ago it became rational for voters to anticipate this outcome of the political process.

Wagner's law, relating taxation to income, has generated a large literature and has been tested in various ways. Our analysis shows that Wagner's law should be amended to include the effect of relative income in addition to absolute income.

Kuznets (1955) observed that economic growth raises the incomes of skilled individuals relative to the incomes of the unskilled. In this way, economic growth can lead to rising inequality and, if our hypothesis is correct, to votes for redistribution. The rising relative size of government slows when the relative changes come to an end and reverses if the relative changes reverse in a mature stationary economy.

The distinctive feature of our analysis is not the voting rule but the relation between individual and collective choice. Each person chooses consumption and leisure by maximizing in the usual way. Those who work receive wages equal to their marginal product. Taxes on labor income provide revenues for redistribution, however, so everyone benefits from decisions to work and incurs a cost when leisure increases.

The analysis explains why the size of government and the tax rate can remain constant yet be criticized by an overwhelming majority of citizens. The reason is that at the voting equilibrium nearly everyone prefers a different outcome. If unconstrained by the voting rule, everyone but the decisive voter would choose a different outcome. But only the decisive voter can ensure a majority.

An extension of our argument may suggest why real government debt per capita, as measured in the budget, has increased more than 20-fold in this century. The decisive voter has as much incentive to tax the future rich as the current rich. An optimal distribution of the cost of redistribution would not tax only the current generation because, with economic growth, the future generation will be richer than the current generation. By shifting the burden of taxation toward the future, income is redistributed intertemporally.

References

Alt, J. (1980). "Democracy and Public Expenditure." Multilithed, Washington University, Political Science Dept., St. Louis, MO.

Aranson, P. and Ordeshook, P. C. (1980). *Alternative Theories of the Growth of Government and Their Implications for Constitutional Tax and Spending Limits.* Multilithed. Carnegie-Mellon University, School of Urban and Public Affairs, Pittsburgh.

Browning, E. K. and Johnson, W. R. (1979). *The Distribution of the Tax Burden.* Washington, D.C.: American Enterprise Institute.

Brunner, K. (1978). "Reflections on the Political Economy of Government: The Persistent Growth of Government." *Schweizerische Zeitschrift Volkwirtschaft und Statis, 114* (Sept.), pp. 649–80.

Buchanan, J. M. and Tullock, G. (1962). *The Calculus of Consent: Logical Foundations of Constitutional Democracy.* Ann Arbor: University of Michigan Press.

Cameron, D. (1978). "The Expansion of the Public Economy: A Comparative Analysis." *American Political Science Review, 72* (Dec.), pp. 1243–61.

Cooter, R. and Helpman, E. (1974). "Optimal Income Taxation for Transfer Payments under Different Social Welfare Criteria." *Quarterly Journal of Economics, 88* (Nov.), pp. 656–70.

Downs, A. (1957). *An Economic Theory of Democracy.* New York: Harper & Row.

Hayek, F. A. (1979). *Law, Legislation and Liberty.* Vol. 3. *The Political Order of a Free People.* Chicago: University of Chicago Press.

Hotelling, H. (1929). "Stability in Competition." *Economic Journal, 39* (Mar.), pp. 41–57.

King, M. A. (1980). "How Effective Have Fiscal Policies Been in Changing the Distribution of Income and Wealth?" *American Economic Review Papers and Procedures, 70* (May), pp. 72–76.

Kuznets, S. (1955). "Economic Growth and Income Inequality." *American Economic Review, 45* (Mar.), pp. 1–28.

Larkey, P. D., Stolp, C., and Winer, M. (1981). "Theorizing about the Growth and Decline of Government: A Research Assessment." *Journal of Public Policy, 1* (May), pp. 157–220.

Meltzer, A. H. and Richard, S. F. (1978). "Why Government Grows (and Grows) in a Democracy." *Public Interest, 52* (Summer), pp. 111–18.

Mueller, D. C. (1976). "Public Choice: A Survey." *Journal of Economic Literature, 14* (June), pp. 395–433.

Musgrave, R. A. (1959). *The Theory of Public Finance: Study in Public Economy.* New York: McGraw-Hill.

Niskanen, W. A. (1971). *Bureaucracy and Representative Government.* Chicago: Aldine-Atherton.

Nutter, G. W. (1978). *Growth of Government in the West.* Washington, D.C.: American Enterprise Institute.

Olson, M. Jr. (1965). *The Logic of Collective Action: Public Goods and the Theory of Groups.* Cambridge, Mass.: Harvard University Press.

Peacock, A. T. (1979). *Economic Analysis of Government and Related Theories.* New York: St. Martin's.

Peacock, A. T. and Wiseman, J. (1961). *The Growth of Public Expenditure in the United Kingdom.* Princeton, N.J.: Princeton University Press.

Pechman, J. A. and Okner, B. A. (1974). *Who Bears the Tax Burden?* Washington, D.C.: Brookings Institute.

Peltzman, S. (1980). "*The Growth of Government.*" *Journal of Law and Economics, 23* (Oct.), pp. 209–87.

Riker, W. H. and Ordeshook, P. C. (1973). *An Introduction to Positive Political Theory.* Englewood Cliffs, N.J.: Prentice-Hall.

Roberts, K. W. S. (1977). "Voting over Income Tax Schedules." *Journal of Public Economics, 8* (Dec.), pp. 329–40.

Romer, T. (1975). "Individual Welfare, Majority Voting, and the Properties of a Linear Income Tax." *Journal of Public Economics, 4* (Feb.), 163–85.

Romer, T. and Rosenthal, H. (1979). "The Elusive Median Voter." *Journal of Public Economics, 12*, pp. 143–70.

Sahota, G. S. (1978) "Theories of Personal Income Distribution: A Survey." *Journal of Economic Literature,* 16 (Mar.), pp. 1–55.

Sheshinski X. (1972). "The Optimal Linear Income Tax." *Review of Economic Studies, 39*, pp. 297–302.

Tocqueville, A. de (1965). *Democracy in America.* New York: Oxford (Reprint of 1835 edition).

Wicksell, K. (1958). *Finanz Theoretische Untersuchungen.* Translated in *Classics in the Theory of Public Finance,* R. A. Musgrave and A. T. Peacock (eds.). London: Macmillan, pp. 72–118.

3

Tests of a Rational Theory of the Size of Government

ALLAN H. MELTZER AND SCOTT F. RICHARD

I. Introduction

The extensive literature on the size and growth of government attests to the long-standing interest of social scientists in the interrelations of economic development, income distribution, political processes, bureaucracy, and tax rates. Surveys of parts of this literature (Cameron, 1978; Larkey, Stolp, & Winer, 1981; Peacock, 1979) show that neither theoretical nor empirical work has resolved the main issues. (See also Peltzman, 1980.) There is little agreement about a common model or framework for predicting the size of government or discussing the causes of governmental growth or decline. And there is no consensus about the empirical evidence on the reasons for changes in the size of government.

The disputes are not about the relevant facts for the twentieth century. Nutter (1978) has shown that if the average size of government is measured by either the share of income taken in taxes or spent by government, the size of government has increased in all developed market economies during the past quarter century. Growth is not a recent phenomenon. Data for the average tax rate or government's share of spending suggest that, judged by these measures, the size of government increased also during the first half of the century in several countries.

Many explanations of the increased size of government emphasize the role of "suppliers" of government serivces. Several studies (Fiorina & Noll, 1978; Niskanen, 1971; Romer and Rosenthal, 1978) suggest that there is an element of

---

We wish to acknowledge the assistance of Joshua Angrist in compiling the data used in our tests, and the helpful advice of Dennis Epple. Thanks also to Peter Aranson for helpful comments and to Messrs. Kilsop and McIntyre of the Central Statistical Office of the United Kingdom for providing data for the tests of U.K. redistribution.

monopoly power on the supply side.[1] Congress, bureaucrats, or "interest groups" are able to raise government spending above the level that utility-maximizing households or voters would choose in the absence of this monopoly power. Models of this kind posit some explicit or implicit cost of acquiring information or taking action and reject the stylized model of the fully informed maximizing voter. Models of "supply" usually exclude the combined effects of taxes and spending on voters' choices by omitting the requirement that the government budget is balanced, in real terms, and other requirements of a general equilibrium solution to the voters' problem.

An earlier tradition placed greater emphasis on voters' demands. De Tocqueville's (1965) perceptive commentaries on early American political arrangements recognize that extension of the franchise to those who do not own property increases the proportion of voters who favor income redistribution. Expressed fears of "mob rule" are an earlier, and less precise, recognition of the same problem. Unbounded fears of this kind have not been realized; governments have increased in size, but their size has been limited. Theories of the demand for government services and redistribution, or outright redistribution of income, that explain the increased size of government should also explain the forces limiting the size of government.

Following Downs (1957), many models of the demand for government services rely on a representative voter who makes the decisive choice (or choices) for society. Roberts (1977) shows that with universal suffrage and majority rule, the median voter is the decisive voter in a specific kind of single-issue election. In our previous work (Meltzer & Richard, 1981), building on Robert's result, we developed a general equilibrium model in which the share of income taken in taxes rises with the difference between mean income and the income of the decisive voter. Franchise extensions and changes in relative productivity alter mean income relative to the income of the decisive voter, and change the tax share. Changes in the proportion of the population receiving social security benefits, not subject to tax, increase votes for higher taxes and redistribution. But higher taxes reduce incentives to work, lower aggregate income, and thus reduce tax collections. All voters recognize both the costs and the benefits of changes in the size of government; the decisive voter chooses the optimum size.

This essay tests our earlier model using annual data for the United States. In the following section we restate the model developed in Meltzer and Richard (1981) using a specific utility function in place of the general function used previously. Our choice of utility function permits us to test for the relation between the level of income and the share of income allocated by government. A positive relation between the two is implied by Wagner's law, one of the most familiar and most frequently tested relations in the literature on the size of government. Our hypothesis qualifies Wagner's law and states the conditions under which it is expected to hold.

Throughout, we identify the size of government with the share of aggregate income allocated by government, but we neglect public goods such as defense,

1. An alternative demand-for-government model developed by Aranson and Ordeshook (1977) treats the political process as a mechanism through which individual demands are revealed.

police protection, veterans' benefits, and highway spending.[2] We also neglect those aspects of the political process conventionally regarded as important for allocation to specific programs. Ours is a theory of the size of the government budget, not of the distribution of the budget among specific programs; our model and empirical tests focus on the share of income redistributed and the reasons that the share is large or small.

II. The Model

The country we consider has a large number of persons who differ in productivity and therefore in earned income. In other respects we assume that people are alike. All have identical tastes, and all treat prices, wages, and tax rates as given, determined by the market process and the political process. There are no monopolies, and there is no collusion, logrolling, or coalition formation. The model is static—there is neither capital nor uncertainty.

The utility function is a Stone–Geary function, linear in the logarithms of the principal arguments. The Stone–Geary function is capable of showing whether the share of income taxed remains constant, increases, or decreases as income changes—that is, whether redistribution of consumption goods is viewed as a "luxury" or a "necessity." Utility depends on consumption, c, and leisure, l.

$$u(c, l) = \ln(c + \gamma) + \alpha \ln(l + \lambda) \quad \lambda > -1 \quad (1)$$

If $\gamma = \lambda = 0$, the utility function specializes to the Cobb–Douglas.

Each person allocates time to labor, n, and to leisure, $l = 1 - n$. Income, y, is measured in units of consumption and is produced using a constant returns-to-scale technology. A person with productivity x earns pretax income y.

$$y = xn(x) \quad (2)$$

The tax rate, t, is a constant fraction of earned income.[3] Tax receipts, tnx, provide r units of consumption per capita; therefore, net redistribution to each household is $r = nxt$.[4] The net payments to government rise monotonically with productivity. At low levels of productivity, net redistribution, $r - nxt$, is positive;

---

2. The appendix describes our classification of government spending as public goods, publicly supplied private goods, and pure redistribution. The appendix also gives nominal values of the data used and a description of sources and methods. We recognize that we have not developed a theory of allocation, so our classifications involve judgment. The data are available on request.

3. By assumption, productivity is not directly observable, so income is taxed instead of productivity.

4. A finding that beneficiaries of particular programs, for example college education, have average income above the population mean does not invalidate our theory. We predict net redistribution and the size of government, not the redistributive effects of particular programs.

the government subsidizes consumption, $c(x)$. There is no saving; consumption equals disposable income—earnings net of taxes plus redistribution.

$$c(x) = (1 - t)nx + r \qquad (3)$$

Each person takes x, t, and r as given and chooses n to maximize utility. Substituting equation (3) and $l = 1 - n$ into equation (1), we obtain

$$\max u(c, l) = \max\{\ln[r + nx(1 - t) + \gamma] + \alpha \ln(1 - n + \lambda)\} \qquad (4)$$

Differentiating (4) and solving the first-order conditions for n gives the labor-leisure choice for those who choose to work and those who choose full-time leisure.

$$n(x) = \begin{cases} \dfrac{(1 + \lambda)x(1 - t) - \alpha(r + \gamma)}{x(1 - t)(1 + \alpha)} & x > x_0 \\ 0 & x \leqslant x_0 \end{cases} \qquad (5)$$

such that

$$x_0 = \frac{\alpha(r + \gamma)}{(1 - t)(1 + \lambda)} \qquad (6)$$

x_0 is the productivity level of the last person who chooses not to work.

If there are neither taxes nor redistribution, $t = r = 0$. In this polar case, n is at a maximum, but $x_0 > 0$ if $\gamma > 0$. (This is a property of the utility function.) From (5) and (6) we see that as r increases, n falls and x_0 rises. Similarly, as the tax rate (t) rises, labor supply, n, falls and x_0 rises.

To find an equilibrium relation between taxes and income that can be estimated empirically, we must show that the system reaches equilibrium for the value of t that the voters choose. This requires three steps. First, we require that consumption and leisure be normal goods, so that increases in disposable income increase consumption and leisure. From (3) and (5) we have $\partial n/\partial r < 0$ and $\partial c/\partial r = 1/(1 + \alpha) > 0$, so these conditions are satisfied. Second, earned income—income before taxes and redistribution—must increase with productivity. If this is not so, we cannot use observed income to infer the relation between votes and productivity. The assumption that leisure and consumption are normal goods implies that

$$\frac{\partial y}{\partial x} = \frac{1 + \lambda}{1 + \alpha} > 0 \quad \text{for } x > x_0 \qquad (7)$$

so the ordering of productivity orders income.

The third step requires a bit more development. We want to show that the choice of the tax rate (or per capita redistribution, r) determines unique equilibrium values for consumption, income, labor (or leisure), and the size of government. Let $F(\cdot)$ be a distribution function for individual productivity, so that $F(x)$ is the fraction of the population with productivity less than x. Per capita income, $\bar{y}$, is given by

$$\bar{y} = \int_{x_0}^{\infty} xn(x)\,dF(x) \tag{8}$$

Rewrite equation (6) as $x_0(1-t)(1+\lambda) = \alpha(r+\gamma)$; substitute this expression into (5), and calculate $y = nx = (1+\lambda)/(1+\alpha)(x - x_0)$, $x \geqslant x_0$. Substituting into (8) yields

$$\bar{y}(x_0) = \frac{1+\lambda}{1+\alpha} \int_{x_0}^{\infty} (x - x_0)\,dF(x) \tag{9}$$

Average income, $\bar{y}$, and therefore aggregate income, is determined once we know the productivity of the last nonworker, x_0. From (6) we see that x_0 depends on the parameters of the utility function and on r and t. The choice of r and t determines equilibrium values of n, y, and c for each person and x_0 for society.

A balanced budget reduces the voter's choice to a single variable, r or t. The requirement that the budget is balanced is not an arbitrary restriction, since r is a real variable and the government (or redistributive mechanism) transfers real resources. Hence the budget must be balanced in real terms. Formally, the budget is balanced when

$$t\bar{y} = r \tag{10}$$

The left side of (10) is a decreasing function of r, since

$$\frac{\partial \bar{y}}{\partial r} = \int_{x_0}^{\infty} x \frac{\partial n}{\partial r} dF(x) < 0$$

Previously we showed that $\partial n/\partial r < 0$. The right side of (10) increases with r. Since the two sides of (10) change in opposite directions as r changes, there is a unique value of r at which the budget is balanced. Equivalently, we can treat (10) as an equation that determines the t that maximizes utility and achieves budget balance for a given r.

The choice of tax rate depends on the relative valuation that the decisive voter places on taxes and redistribution, and therefore on the productivity of the decisive voter. If the decisive voter has $x \leqslant x_0$, the voter does not work. The voter's preferred choice is for maximum redistribution, but maximum redistribution is obtained at $t < 1$ as long as some people respond to incentives to work. In this analysis, $\partial \bar{y}/\partial r < 0$ is sufficient to ensure that the maximum tax is below $t = 1$. A decisive voter with above-average productivity recognizes that his net taxes are positive at any $t > 0$, so he chooses $t = 0$. A decisive voter between these extremes balances the utility gain from increased redistribution against the utility loss from higher taxes. In reaching a decision, the voter is concerned only with his own utility, but his utility depends on everyone's decision to work, because all taxpayers contribute to the budget and the amount available for redistribution.[5]

Formally, the decisive voter chooses a tax rate that balances the budget and maximizes utility, equation (4), subject to his budget constraint, equation (3). Let x_d and n_d denote the productivity and hours of work for the decisive voter; $y_d = x_d n_d$ is then the earned income of the decisive voter. The preferred tax rate is the rate t that satisfies[6]

$$\bar{y} + t\frac{d\bar{y}}{dt} - y_d = 0 \tag{11}$$

By differentiating (9) and using (6) and (10), we find that[7]

$$\frac{d\bar{y}}{dt} = -\frac{\alpha[1 - F(x_0)](\bar{y} + \gamma)}{(1-t)\{1 + \alpha - t[1 + \alpha F(x_0)]\}} \tag{12}$$

---

5. Everyone who works receives a wage equal to his marginal (and average) productivity, as can be seen from (2), and pays a tax.

6. To derive (11), substitute (10) and $y_d = n_d x_d$ into (4), to get

$$u = \ln[t\bar{y} + y_d(1-t) + \gamma] + \alpha \ln(1 - n_d + \lambda)$$

Then

$$\frac{du}{dt} = \frac{\bar{y} + t(d\bar{y})/(dt) - y_d}{t\bar{y} + y_d(1-t) + \gamma} + \frac{dn_d}{dt}\left[\frac{x_d(1-t)}{t\bar{y} + y_d(1-t) + \gamma} - \frac{\alpha}{1 - n_d + \lambda}\right]$$

The second term is zero for a nonworker, since $dn_d/dt = 0$; it is also zero for a worker, as can be seen by substituting (5) into the coefficient of dn_d/dt:

$$\text{coeff of } \frac{dn_d}{dt} = \frac{x_d(1-t)(1+\lambda-n_d) - \alpha[r + \gamma + x_d n_d(1-t)]}{D}$$

$$= \frac{1}{D[x_d(1-t)(1+\lambda) - x_d n_d(1-t)(1+\alpha) - \alpha(r+\gamma)]}$$

$$= \frac{1}{D[x_d(1-t)(1+\lambda) - (1+\lambda)x_d(1-t) + \alpha(r+\gamma) - \alpha(r+\gamma)]} = 0$$

where $D = [r + \gamma + x_d n_d(1-t)](1 + \lambda - n_d)$. Utility maximization requires that $du/dt = 0$, which implies (11).

7. Differentiating (9) we get

$$\frac{d\bar{y}}{dt} = \frac{1+\lambda}{1+\alpha}\int_{x_0}^{\infty}\frac{dx_0}{dt}dF(x) = -\frac{1+\lambda}{1+\alpha}\frac{dx_0}{dt}[1 - F(x_0)] \tag{F1}$$

Differentiating (6) we find

$$\frac{dx_0}{dt} = \frac{\alpha[(dr)/(dt)(1-t) + r + \gamma]}{(1+\lambda)(1-t)^2} \tag{F2}$$

such that $F(x_0)$ is the fraction of the population that does not work. Substituting (12) into (11) and denoting $g = \gamma/y_d$ and $m = \bar{y}/y_d$, we find

$$0 = \frac{1 + \alpha F(x_0)}{\alpha[1 - F(x_0)]}(m - 1)(1 - t)^2 + (2m + g - 1)(1 - t) - (m + g) \qquad (13)$$

The solution to equation (13) is the optimal tax rate for a decisive voter who works and chooses $t > 0$.

Equation (13) is an equilibrium relation between t, m, and F. For given productivity and tastes, the decisive voter's choice of t determines x_0, $\bar{y}$, and m, and all other endogenous variables follow. When making his choice, the decisive voter is aware that he cannot treat (13) as a quadratic function in t. The reason is that m and x_0 depend on the choice of t.

III. Empirical Tests

To estimate equation (13) we take a linear approximation, specify a decisive voter, and choose empirical counterparts for m, $F(x_0)$, g, and t. This section discusses the choices—and compromises—we make, derives the expected signs of coefficients, and presents the results of our estimation.

Roberts (1977) shows that the voter with median income is decisive in single-issue elections to choose a linear tax rate. His results do not consider the effect of turnout.[8] In practice, differential turnout and restrictions on voting drive a wedge between the person with median income and the median voter defined in voting models. We cannot locate comprehensive time series data on the distribution of voters' earned income, so we use the income of the median income earner as a proxy for the income of the median voter. We assume that the person with median income is decisive in elections to choose the tax rate, so m is the ratio of mean to median income and g is the ratio of γ to median income.

Now $r = t\bar{y}$ can be differentiated to yield

$$\frac{dr}{dt} = \bar{y} + t\frac{d\bar{y}}{dt} \qquad (F3)$$

Substituting (F3) and $r = t\bar{y}$ into (F2) gives

$$\frac{dx_0}{dt} = \frac{\alpha\{[\bar{y} + t(d\bar{y})/(dt)](1 - t) + t\bar{y} + \gamma\}}{(1 + \lambda)(1 - t)^2} \qquad (F4)$$

Finally we substitute (F4) into (F1) and collect the term in $d\bar{y}/dt$ to get (12).

8. Several papers show that median voter theorems require strong restrictions on tastes. A summary of this literature is provided by Aranson and Ordeshook (1981). Empirical work yields ambiguous conclusions. Romer and Rosenthal (1979) show that many of the tests do not distinguish between the median and other fractiles, so no firm conclusion can be drawn from them. Cooter and Helpman (1974) estimate the shape of the social welfare function implicit in U.S. data and conclude that the data are consistent with the median voter rule.

To obtain a linear approximation of (13) we solve for $1 - t$.

$$1-t=\frac{-2m+1-g+(1+g)\{1+(4b/1+g)(m-1)+[4b/(1+g)^2](m-1)^2\}^{1/2}}{2(m-1)(b-1)}$$

such that $b = (1 + \alpha)/\alpha(1 - F)$. Expanding by means of a first-order approximation in $m - 1$ and g gives as an approximation[9]

$$t \approx \left(\frac{1+\alpha}{\alpha}\right)\frac{m-1}{1-F}\frac{1}{1+g}$$

Taking logs and letting g approximate $\ln(1 + g)$ for small g, we have

$$\ln t + \ln(1 - F) = \ln\frac{1+\alpha}{\alpha} + \ln(m-1) - \frac{\gamma}{y_d} \qquad (14)$$

Two problems remain in estimating (14). The first is the possibility of simultaneous-equations bias. This possibility arises because (14) is an equilibrium relation containing t and $F(x_0)$, and the same process determines both. We avoid this problem by using $\ln t + \ln(1 - F)$ as the dependent variable.[10] The second problem is the specification of a measure of the distribution of income that is not subject to large reporting bias and is available for a period long enough to induce a noticeable change in the size of government. We use data from social security payments to compute m as the ratio of mean to median income from labor services and to measure y_d. War years are omitted. $F(x_0)$ is measured by the proportion of the population receiving payments for old age, disability, blindness, and aid to families with dependent children. The share of earned income used for redistribution (t) is computed from federal, state, and local data. (The appendix describes the data and our procedures more fully.) Public goods are omitted.

Table 3-1 shows the estimates of the parameters of equation (14). The data generally support the hypothesis, but we must note some reservations. Equation (14) implies that the coefficient of $\ln(m - 1)$ is plus one. We obtain the expected sign, but the coefficient is much below unity. Some of the problem undoubtedly results from our use of a linear approximation. The derivative of t with respect to m from (13) is not constant and bears no obvious relation to unity. Further, the estimated elasticity of $t(1 - F)$ with respect to $m - 1$, 0.48, implies that, at

9. We use the following approximation for small x:

$$(1 + px + qx^2)^{1/2} \approx 1 + \frac{p}{2x} + \frac{1}{2}\left[q - \left(\frac{p^2}{4}\right)\right]x^2 + \sigma(x^3)$$

10. An additional aspect of this problem arises from our use of $m - 1$ and y_d as independent variables. To avoid this problem, we require a theory of the (simultaneously determined) distribution of income, a task far beyond this chapter. We chose to treat $m - 1$ and y_d as determinants of the levels of t and x_0 for our preliminary test. The converse has little appeal to us.

Table 3-1. Least squares estimates of equation (14), using data from 1937–40 and 1946–76[a]

Dependent Variable	$\ln(m-1)$	$1/y_d$	Constant	R^2	DW[b]	HL[c]
$\ln t(1-F)$	0.57 (9.1)[d]	−1081 (5.0)	−1.09	0.80	1.30	—
$\ln t(1-F)$	0.48 (5.5)	−1402 (4.3)	−1.13	0.81	1.61[e]	0.3
$\ln t_2(1-F)$	0.48 (9.2)	28.3 (0.16)	−1.95	0.73	1.09	—
$\ln t_2(1-F)$	0.34 (6.0)	−219 (1.0)	−2.08	0.80	1.72[e]	0.3
$\ln t_3(1-F)$	0.67 (5.5)	−3461 (8.1)	−1.37	0.79	1.71[e]	—
$\ln t_3(1-F)$	0.71 (4.0)	−3781 (5.8)	−1.25	0.79	1.90[e]	0.3

[a] See text and appendix for description of data and procedures.
[b] DW is the Durbin–Watson statistic.
[c] HL indicates the value of ρ in the Hildreth–Lu procedure. HL correction loses two observations, 1937 and 1946.
[d] t Statistics are shown in parentheses. $t = t_2 + t_3$ is the sum of publicly supplied private goods (t_2) and "pure" redistribution (t_3).
[e] Denotes that serial correlation is rejected or indeterminate at $\rho = 0.05$ level for the value of ρ shown under HL using a one-tailed test.

current values of net national product, a 1-percent change in $m-1$, with $F(x_0)$ held constant, changes total spending on redistribution by $1.5 billion. A $3-billion increase, implied by a unit coefficient on $\ln(m-1)$, seems too large, suggesting that the linear approximation overstates the response.[11]

Wagner's law is one of the oldest and most frequently tested explanations of the growth of government. This law has been interpreted in two ways. The traditional interpretation is that government services are a luxury good, so a positive relation exists between the relative size of government and the level of real income. Alt (1980, p. 4) has questioned this interpretation of Wagner's idea. He notes that Wagner stated that there is "a proportion between public expenditure and national income which may not be permanently overstepped." This suggests an equilibrium relative size of government rather than an ever-growing government sector.

The traditional statement of Wagner's law—that government grows more rapidly than income—has been tested many times, with mixed results. Relatively recent surveys discuss tests of Wagner's law applied to different areas of government activity (Cameron, 1978; Larkey, Stolp, & Winer, 1981).

Our model suggests that previous tests of Wagner's law may suffer for two reasons. First, they omit a key explanatory variable—some measure of the relative income distribution. Second, they use aggregate income or average

---

11. Regressions with $1-F$ on the right side of the equation produce the expected signs for $\ln(m-1)$ and $\ln(1-F)$, but the coefficient differs from unity, the value implied by (14). The use of a linear approximation may explain some of the difference between implied and computed values.

Table 3-2. The effect of aggregate income[a]

Dependent Variable	$\ln(m-1)$	$\ln y$	Constant	R^2	DW	HL
$\ln t(1-F)$	.51	.19	−2.75	.81	1.31	—
	(7.8)	(5.2)				
$\ln t(1-F)$	.35	.25	−3.33	.83	1.67[e]	.3
	(4.0)	(5.0)				
$\ln t_2(1-F)$	.49	−.01	−1.87	.74	1.10	—
	(8.8)	(.30)				
$\ln t_2(1-F)$	.31	.05	−2.53	.81	1.70[e]	.3
	(5.2)	(1.6)				
$\ln t_3(1-F)$	.46	.62	−6.69	.82	1.62[e]	—
	(3.8)	(9.1)				
$\ln t_3(1-F)$	.38	.64	−6.89	.80	1.93[e]	.3
	(2.1)	(6.0)				

[a] See footnotes a–d to Table 3-1. y is real GNP.
[e] Denotes that serial correlation is rejected or indeterminate at $p = .05$ level for the value of ρ shown under HL using a one-tailed test.

income as an explanatory variable instead of the reciprocal of median income. Equation (14) implies a relation between the ratio of mean to median income, m, and the share of income allocated by government, but it also implies a relation between the reciprocal of median income and the share of income allocated by government. The direction of the effect and its existence depend on γ. When $\gamma = 0$, the size of government is independent of the level of income, because the marginal utility of consumption is independent of income. A positive value of γ lowers the marginal utility of consumption at each level of income. The decisive voter, with $\gamma > 0$, votes for higher taxes and more redistribution despite the reduction in aggregate income and consumption that results. A negative value of γ implies that taxes and redistribution decline as median income rises.

The use of aggregate income instead of the reciprocal of median income is not a serious source of misspecification in our test of Wagner's law. Table 3-2 shows that if we substitute the log of real GNP for $1/y_d$, the qualitative results remain unaffected. Therefore we may treat our result as a test of Wagner's law and our findings as evidence that Wagner's law must be amended to include the effect of relative income.

The findings support three conclusions about redistribution. First, the share of income redistributed increases with income, but the size of the response depends on the level of (median) income. The reason is that the two components of redistribution that we study—"pure" redistribution (t_3) and publicly supplied private goods (t_2)—respond differently to income.[12] Income has a considerably larger and statistically more reliable effect on t_3 than on t_2. Increases in real

---

12. We do not provide a theory of the distribution of expenditures between t_2 and t_3, so our classification is judgmental. We did not try to reclassify between t_2, t_3, and public goods, so we can only conjecture about the extent to which reclassification would affect our results.

income appear to stimulate demand for t_3 and t, and reductions of income appear to reduce the demand for t_3 and t but do not affect t_2.

Second, Wagner's law is not a general law but must be qualified. The conflicting evidence for and against various versions of Wagner's law, found in previous studies, may reflect a failure to allow for changes in the composition of redistribution. As income rises, people choose relatively more redistribution in cash, which permits maintenance of consumption with less work. The growth of social security transfers and unemployment compensation, major programs redistributing cash, is consistent with this finding.

Third, the results generally support our hypothesis and suggest that a substantial part of the growth of government is a response to voter demand. Although we have not tested, and cannot exclude, other plausible explanations—including the much discussed pressure groups, bureaucrats, and politicians—our findings support an explanation based on rational choice, and assign a nonnegligible weight to voters' choice of the size of government as an explanation of observed changes in the relative size of government.

III. Conclusion

In many countries the political party holding power changes more frequently than the trend growth rates in government spending and taxes. Shifts of political power are often preceded by rhetoric about "meeting needs" or cutting taxes, which postelection performances rarely match. Yet changes in the size of government, up and down, occur. On average, the size of government, measured by the ratio of the share of taxes or spending to output, has increased in many countries during the past century. But changes in size occur at different rates, and at times the relative size of government declines.

Meltzer & Richard, (1981) develops a general equilibrium model in which people choose consumption and leisure and, as voters, decide on income redistribution or the (average) tax rate. The model implies that the size of government changes with the ratio of mean income to the income of the decisive voter and with the voting rule or qualifications for voting. Extensions of the franchise that increase the number of voters who benefit from income redistribution increase votes for redistribution. The size of government increases. Changes in the age composition of the population that increase the proportion of the population receiving old age assistance also increase redistribution paid from taxes on labor income. The relation is symmetric. Changes in productivity, or in labor force participation, that lower mean income relative to the income of the decisive voter reduce the size of government.

The tests of the model reported here treat the person with median income as the decisive voter. We find that the ratio of government spending for redistribution to aggregate income, and the share of aggregate income redistributed in cash, rise and fall with the ratio of mean to median income and the level of (median) income. Redistribution in kind—the provision of education, health care, fire protection, and other services—also rises and falls with the ratio of mean to median income, but it appears to be independent of the level of income.

Our model implies, and the data suggest, that Wagner's law—relating the size of government to the level of income—must be amended. The relation is not simple and direct, as many tests of Wagner's law presuppose, but depends, in our model, on shape of the income distribution—more specifically on the ratio of mean to median income. Failure to hold the distribution of income constant renders many previous tests meaningless. Further, our results suggest that voters' choice of the nature of redistribution, in cash or in kind, affects the results. Although we do not derive the relation between size of government and voters' choice of cash or in-kind benefits (and we neglect public goods), our results suggest that Wagner's law is more likely to find support if redistribution is in cash.

Our hypothesis is parsimonious: There is no uncertainty; taxes are linear; all redistribution is by lump sum transfer; and the decisive voter is fully cognizant of the costs and benefits of the redistribution he demands, including the effects on incentives to work and consume. We neglect most features of the political process, including any influence of interest groups, bureaucrats, and other monopoly elements that affect "supply." We recognize that a useful extension of our model would incorporate the allocation of funds to specific programs and thereby incorporate supply factors. In our empirical work, we rely on a linear approximation to a nonlinear equation and obtain our estimates from an equilibrium relation, not a structural equation. Despite the model's parsimony, the neglect of supply factors, the use of a linear approximation, and data interpolation, the hypothesis explains much of the trend in the relative size of spending for redistribution and a considerable part of the annual variation observed in the United States during a recent 40-year period.

During the period we studied, our measure of government spending for redistribution increased, in nominal value, from $10 billion to more than $350 billion and the share of total income redistributed rose from 12 to 22 percent. A considerable part of the observed increase in the size of government appears to be consistent with rational choice by maximizing voters who benefit from redistribution and are able to shift a disproportionate share of the cost to people with incomes above the mean.

IV. Postscript 1990

Shortly after the tests were published, we obtained data for the United Kingdom that permitted additional tests of the hypothesis to be carried out.[13] Table 3-3 shows some of the estimates. Definitions and sources of data are found in the 1990 addendum to the data appendix.

---

13. We are very much indebted to Joshua Angrist. While an undergraduate student in London, Angrist persuaded the statisticians at the Inland Revenue to release unpublished data on the ratio of mean to median income to retest our hypothesis, and he collected the other data required for the estimation.

Table 3-3.[a]

Dependent Variable	$\ln(m-1)$	$1/y_d$	Dummy 1975	Constant	R^2	DW
$\ln t(1-F)$	0.17 (1.78)	−12.9 (16.4)	0.09 (2.23)	−0.51 (3.29)	0.94	1.62
$\ln t_2(1-F)$	−0.01 (0.10)	−13.4 (11.8)	0.07 (1.14)	−1.44 (6.44)	0.88	1.48
$\ln t_3(1-F)$	0.34 (2.35)	−12.5 (10.2)	0.12 (1.85)	0.97 (4.07)	0.87	1.54

[a]See footnotes to Table 3-1. Estimation period is 1953–1979. Variables are defined differently than in the United States. See appendix.

Table 3-4.[a]

Dependent Variable	$\ln(m-1)$	$1/y_d$	Dummy 1975	Constant	R^2	DW
$\ln(S/Y)(1-F_2)$	0.32 (2.49)	−16.6 (15.5)	0.08 (1.33)	−0.84 (4.01)	0.93	1.53

[a]See footnotes to Table 3-1. Estimation period is 1953–1979.

The responses of redistributive expenditures to $\ln(m-1)$ is smaller in the U.K. data than in the U.S. data; they are significant in the t_3 (pure redistribution) equation but not in the t_2 (the public supply of private goods). It seems clear that our measure of income provides much of the explanatory power of the equation, in contrast to the U.S. data.[14]

We also estimated the response of current grants and subsidies to the variables in our model. Grants and subsidies include payments to individuals, mainly national insurance benefits, and subsidy payments to business enterprises in the public and private sectors, made by local and central governments. This value is divided by national income at factor cost. It is shown in Table 3-4 as S/Y multiplied by $1-F_2$, where F_2 is the share of population receiving family allowances. Again, the data support the hypothesis for this alternative measure of each transfer.

Data Appendix

I. Main Sources of Data
    A. Government Spending. Fiscal years 1936–72: U.S. Bureau of the Census, *Census of Governments*, Vol. 6, no. 4. "Historical Statistics on

---

14. The dummy for 1975 picks up discontinuities arising from changes in the definition of income in that year.

Government Finances and Unemployment," Tables 1, 3, and 4. Fiscal Years 1972–77: U.S. Bureau of the Census, *Census of Governments, 1977*. Data were available for all years 1952–77 and for even years 1936–50. Three sources were used to interpolate the odd-numbered years 1937 to 1951 inclusive.
1. The Budget of the U.S. Government, summary budget statements and "Message of the President" for fiscal years 1946, 1948, 1952, and 1955
2. Bureau of the Census, *Compendium of State Government Finances*. (Also published as *Financial Statistics of States* [year] and as *State Finances* [year])
3. Bureau of the Census, *Compendium of City Government Finances* (year)

B. Welfare recipients, including old age, disability, blindness, and families receiving aid to families with dependent children. 1937–49: *Annual Statistical Supplement to Social Security Bulletin*, Table 60, "Public assistance and federal work programs," various years. 1950–77: *Statistical Abstract of the U.S.*, Table 562, "Public aid," various years.

C. Mean and Median income: 1937–76: *Social Security Bulletin, Annual Statistical Supplement*, various years, for example 1974, Table 37, for median earnings all workers, and Table 34, for mean earnings. 1949–64: Bureau of the Census, Technical Paper 17; 1964–76: "Current Population Reports," Series P. 60.

D. Gross national product 1937–38: *Economic Report of the President*, Jan, 1971. 1939–76: *Economic Report of the President*, Jan. 1980.

II. Definitions of Government Spending
  A. Public supply of private goods includes
     1. postal service
     2. higher education
     3. local schools
     4. other education
     5. hospitals
     6. local fire
     7. local sanitation
     8. other natural resources
     9. nonhighway transportation
     10. utilities and liquor stores
  B. Redistribution includes
     1. public assistance
     2. public welfare
     3. stabilization of farm prices and incomes
     4. housing and urban renewal
     5. unemployment compensation
     6. old age, disability, and survivors insurance
     7. other insurance

C. Public goods: remaining items, mainly police, defense, veterans benefits, and highways. These items are *not* used in our statistical estimation.
D. Remaining items include "other and unallocable" and employee retirement.

III. Interpolation and Adjustment
A. General interpolation for odd-numbered years between 1936 and 1952. Data for state and local spending was not available for these odd-numbered years. Data were taken from the *Budget of the United States Government*, summary budget statements, the Census Bureau's annual *Compendium of State Government Finances* and the annual *Compendium of City Government Finances*. Data were deflated by an index of wages for full-time equivalent employees in federal, state, and local government using data from the national income account (*National Income and Product Accounts of the United States, 1927–74*, p. 210). Public supply of private goods and redistribution were treated separately. A linear regression was run using the deflated data as dependent variable and the longer, more comprehensive time series from the *Census of Governments* as independent variables. Observations for the eight *even* years 1938–52 inclusive were used to estimate coefficients, and the coefficients were used with data based on the *Compendium of State Government Finances* and the *Compendium of City Government Finances* to interpolate the odd-numbered years 1937–1951 inclusive that were not available in the *Census of Governments* data base. The interpolated data were then transformed to nominal terms. The proportion of the series adjusted using the nominal wage of federal employees was computed using the share of federal expenditure in the unadjusted series for both major categories of spending. Data for war years 1941–45 inclusive were computed but not used.
B. Specific adjustments: public supply of private goods. Merchant marine expenditure was highly correlated with war-time spending. These data were removed from the series used in the regression. Data were added back to the interpolated series. Data for utility and liquor stores could not be obtained from *Compendium* with sufficient comparability to data in *Census of Governments*. Linear interpolations of the *Census of Governments* were used for odd-numbered years between 1936 and 1952.
C. Shares of income redistributed were computed by deflating spending by fiscal year nominal net national product.

Addendum on U.K. Data 1990

I. Mean and Median Income

There are two sources of data on U.K. mean and median income, one from the Inland Revenue and one from household surveys. We were advised that Inland Revenue data are more accurate, so we report estimates based on these data.

The series are mean and median incomes for tax units before income tax. A tax unit is either an unmarried individual or a married couple. A unit includes people below voting age.

The definition of income was revised in 1975. We used a dummy variable to estimate the shift, if any, in response. Prior to 1968, data are for calendar years. For 1968 and after, data are for fiscal years. Income includes some transfers from public authorities.

II. Public Supply of Private Goods (t_2)

This series includes the following expenditure items: employment serivces; payments for agriculture, fisheries, and food; research; water refuse and sewage; libraries, museums, and arts; fire services; education; national health; records, registration, and surveys; transport and communication; miscellaneous local government services.

III. Redistribution (t_3)

The series includes payments for: housing; local welfare and child welfare; school meals, milk, and food; personal social services; social security benefits. War pensions are excluded.

IV. The Spending Share $t = t_2 + t_3$

The component time series were compiled from different editions of statistics on national income and expenditure, and they are not adjusted fully for changes in definition and data revisions.

The series were deflated by gross domestic product at factor cost. This series is from *Economic Trends*, Supplement 1981, Table 151.

V. The Proportion of Nonworkers $(1 - F)$

F is measured as the number of persons receiving retirement pensions divided by population. Series are from *Annual Abstract of Statistics* 1965, 1968, 1976, 1981. Population, same source, 1979.

VI. Subsidies and Grants (S)

Data are from *Economic Trends*, 1981, Table 151.

References

Alt, J. (1980). Democracy and Public Expenditure. Multilith, Washington University, St. Louis, Political Science Department.

Aranson, P. H. and Ordeshook, P. C. (1977). *"Incrementalism, the Fiscal Illusion, and the Growth of Government in Representative Democracies."* Multilith, Fourth Interlaken Seminar, June. Emory Univ., Economics Department.

Aranson, P. H. and Ordeshook, P. C. (1981). "Regulation, Redistribution, and Public Choice." *Carnegie Papers on Political Economy*, a supplement to *Public Choice, 37*, pp. 69–100.

Cameron, D. (1978). "The Expansion of the Public Economy: A Comparative Analysis." *American Political Science Review, 72* (Dec.), pp. 1243–61.

Cooter, R. and Helpman, E. (1974). "Optimal Income Taxation for Transfer Payments under Different Social Welfare Criteria." *Quarterly Journal of Economics, 88* (Nov.), pp. 656–70.

Downs, A. (1957). *An Economic Theory of Democracy*. New York: Harper and Row.

Fiorina, M. P. and Noll, R. G. (1978). "Voters, Bureaucrats, and Legislators: A Rational Choice Perspective on the Growth of Bureaucracy." *Journal of Public Economics, 7*, pp. 239–54.

Larkey, P. D., Stolp, C., and Winer, M. (1981). "Theorizing about the Growth and Decline of Government: A Research Assessment." *Journal of Public Policy, 1* (May), pp. 157–220.

Meltzer, A. H. and Richard, S. F. (1981). "A Rational Theory of the Size of Government." *Journal of Political Economy, 89* (Oct.), pp. 914–27.

Niskanen, W. A. (1971). *Bureaucracy and Representative Government*. Chicago: Aldine-Atherton.

Nutter, G. W. (1978). *Growth of Government in the West*. Washington, D.C.: American Enterprise Institute.

Peacock, A. (1979). *The Economic Analysis of Government and Related Theories*. New York: St. Martin's Press.

Peltzman, S. (1980). "The Growth of Government." *Journal of Law and Economics, 23* (Oct.), pp. 209–87.

Roberts, K. W. S. (1977). "Voting over Income Tax Schedules." *Journal of Public Economics, 8*, pp. 329–40.

Romer, T. and Rosenthal, H. (1978). "Political Resource Allocation, Controlled Agendas and the Status Quo." *Public Choice, 33* (4), pp. 27–43.

Romer, T. and Rosenthal, H. (1979). "The Elusive Median Voter." *Journal of Public Economics, 12*, pp. 143–70.

de Tocqueville, A. (1965). *Democracy in America*. New York: Oxford. (Reprint of the 1835 edition.)

4

A Positive Theory of In-Kind Transfers and the Negative Income Tax

ALLAN H. MELTZER AND SCOTT F. RICHARD

I. Introduction

Few welfare-policy proposals have had greater acceptance among economists and policy advisers in recent years than the proposals for some form of negative income tax. Three presidents—Presidents Nixon, Ford, and Carter—have proposed a variant of the negative income tax to Congress. Social scientists, who hold divergent views on many issues, have favored a negative income tax as either a partial or total substitute for current redistribution policies (see Friedman, 1962; Tobin, 1966; Tobin, Pechman, & Mieszkowski, 1967). Yet not a single country had adopted such a scheme by 1980,[1] and Friedman, who first proposed the scheme, stated that a comprehensive negative income tax has "no chance whatsoever of being enacted at present" (Friedman and Friedman, 1980, p. 124).[2]

Friedman's argument for the negative income tax is probably the most frequently cited argument in favor of cash payments for income redistribution. Friedman makes three claims (Friedman, 1962, pp. 192–93). First is the familiar economic principle that individual welfare increases when the consumer is allowed to choose a preferred consumption bundle. Second, Friedman recognizes that a recipient's incentive to work is reduced by any type of transfer but, he argues, the negative income tax does not eliminate the incentive to work because part of any additional dollars of earned income can be spent for present

---

An earlier version of this chapter was presented at the Interlaken Seminar in 1980. We are indebted to the seminar participants and to Thomas Borcherding, Thomas Romer, and Howard Rosenthal for helpful comments.

1. A letter from Robert Bedand of the International Social Security Association, dated October 22, 1980, advised that "no country has yet, to our knowledge, actually voted any negative income tax programme into law."

2. Friedman and Friedman's argument about political infeasibility relies on Chapter 6 of Martin Anderson's explanation of the failure of comprehensive welfare reform. See Anderson (1978).

or future consumption. Third, the cost of administering the system is reduced, so taxpayers deliver redistribution at lower cost.

Friedman's argument neglects the incentives faced by Congress and the voters who elect its members. Congress, or the Parliament, must agree to the amount and type of redistribution.[3] Anderson (1978) argues that, to be approved by Congress, a comprehensive redistribution program must provide "a decent level of support for those on welfare, ... strong incentives to work, ... [at] reasonable cost" (p. 135). Despite the ambiguity introduced by the modifiers, Anderson's criteria are clear: the size of benefits, costs, and work incentives.[4]

The negative income tax meets only two of Anderson's criteria: It reduces the administrative and monitoring costs of providing a given level of benefits. The problem is with incentives to work. Anderson estimated that, in the early 1970s, recipients of the proposed negative income tax would have faced a marginal tax rate above 50 percent and perhaps above 70 percent if total spending on benefits remained approximately equal to the benefits available under existing programs.[5]

Positive analysis of the negative income tax is usually restricted to calculation of the combinations of benefits, marginal tax rates, and minimum income level or the trade-offs among the three (see Tobin et al., 1967, for example.) Most policy discussion is normative; it does not show who gains and who loses from the adoption of a negative income tax or why rational voters would adopt the plan. This chapter attempts to fill the gap by analyzing the conditions under which cash or cash-equivalent transfers are chosen or rejected by voters constrained to vote on this issue in a single-issue election.

In our model, utility-maximizing people vote on the mix of goods to be redistributed. They can choose to distribute the mix of goods that the recipients prefer most. This is equivalent to redistributing general purchasing power or cash, so we call it the *cash-equivalent transfer*. As in our previous work (Meltzer & Richard, 1981, 1983), the choice is made under a voting rule that permits a decisive voter to make binding choices of society's tax rates and the amount and type of redistribution. To avoid making decisions depend on nonobservable differences in taste, we assume that individuals have identical utility functions. Their endowments differ, however, because individual labor productivity differs. Productivity determines income and, through the household budget constraint, determines consumption.

People consume two traded goods and leisure. A separate linear consumption tax is levied against the purchase of each good. Voters can subsidize purchases of one good by choosing a negative tax rate. Other combinations of

3. See Pommerehne (1979) for an earlier statement of this view. Pommerehne follows Hochman and Rodgers (1969) by including the recipient's utility in the payer's utility function.

4. Friedman and Friedman (1980, pp. 125–26) cite this passage approvingly. They then emphasize the current level of benefits as the main political obstacle.

5. See Anderson (1978, pp. 142–45) for the conflicts of objectives and high marginal tax rates in the Nixon administration's Family Assistance Plan. See also Tobin (1966) for an early statement of the conflict.

linear rates or subsidies—including income taxes, sales taxes, or taxes on wage payments—are special cases of consumption taxes. In particular, when the two tax rates are equal, the consumption tax is equivalent to an income tax.

Tax receipts are redistributed as lump-sum transfers, as in Romer (1975) and Meltzer and Richard (1981), but now, with two consumption goods, the mix of goods becomes a key policy variable. Decisive voters can redistribute any mix of goods they please, including the cash-equivalent mix. They also may set taxes or subsidies as they please subject to the constraint that the government's real budget is balanced. If they have chosen equal tax rates and the cash-equivalent mix, they have in effect chosen a negative income tax. Their decision depends on the voting rule, on the distribution of productivity, and on their preferences.

Redistribution in kind increases incentives to work by restricting the choice of goods and services. An example suggests the reasoning. People with relatively low productivity who receive cash can buy the basket of goods and services they prefer. If, instead of cash, they are offered housing allowances, medical services, and education of equivalent value, they are induced to work to buy other consumption goods. By limiting opportunities to resell goods received as in-kind transfers, people are encouraged to substitute labor for leisure. In-kind transfers and restrictions on resale induce the behavior that taxpayers prefer: more work by those who depend on redistribution.[6]

Decisive voters also decide whether to tax both goods at the same rate or each at a separate rate. The former is a type of direct tax; the latter is an indirect tax. It is in decisive voters' interest to tax most heavily marginal purchases of the goods they redistribute and to subsidize, or tax less, the goods they purchase most heavily. Since all voters share the same tastes but have different incomes, and purchases increase with income, indirect taxes on the good distributed fall most heavily on people with highest incomes. Equal taxation of all goods removes this opportunity to raise taxes on the "rich." For this reason, decisive voters may rely on indirect taxes to finance transfers.

By considering the choice of indirect or direct taxes as part of the decision about income redistribution, our analysis puts more attention on what is usually called vertical equity. Earlier work, beginning with Ramsey's (1927) classic article, analyzed the choice of direct and indirect taxes as a problem of efficiency. Notable contributions by Little (1951), Corlett and Hague (1953), and Harberger (1964) removed some of the restrictions of the Ramsey analysis by introducing labor and leisure and by showing that efficiency losses are minimized if indirect taxes are highest on the good that is complementary with leisure. This result follows from the fact that leisure is not taxed. Higher taxes on complements of leisure are an indirect way of taxing leisure. These results remain when individuals differ, as in the models of Diamond and Mirrlees (1971) and Atkinson and Stiglitz (1980, Chap. 11), but the emphasis shifts toward redistribution. Our analysis differs from this earlier literature by using a

---

6. The size of the effect depends on the elasticity of labor supply in the relevant range. Anderson (1978, p. 100) summarize eight studies of the effect of an income guarantee. If we treat the in-kind transfer as a guarantee of a smaller real income than the equivalent cash transfer, the relevant average elasticity is -0.3.

simplified model of voting, making redistribution a political-economic decision, instead of relying on a social-welfare function to determine tax policy, and by linking the choice of taxes and the choice of spending in the analysis of income redistribution through the government budget.[7]

Although a negative income tax has not been adopted, income redistribution and transfer payments often include both cash payments and in-kind transfers. Unemployment compensation and social security benefits are examples of cash payments that coexist with in-kind transfers. Our analysis yields some testable implications about the conditions under which cash transfers are made and the conditions under which such transfers are avoided.

In the following section we set out a model of individual choice and our assumptions about taxing and spending. Section III derives the decisive voter's choice of taxes and redistribution. The choice depends on the decisive voter and therefore depends on the voting rule. In Section IV we examine the special case of an income tax and the conditions for a negative income tax. The final section states our conclusions.

II. The Model Economy

We consider a simple model of a static economy in which redistribution in kind has a role. Each individual, or household, receives redistribution in goods. Each maximizes utility by dividing time between labor and leisure and by consuming the two goods. Those who work purchase and consume additional goods, but some subsist on transfers. A particular transfer, called the cash-equivalent transfer, is the mix of goods that consumers would buy it all transfers were made in cash.

Goods cannot be resold. This restriction prevents *some* individuals from reaching the welfare maximizing mix that we call the *cash-equivalent mix*.[8] Because we do not permit resale, the redistribution of goods in kind constrains individuals away from consumption bundles they prefer. The use of in-kind transfers in turn affects the trade-offs individuals make between labor and leisure. The decisive voter exploits these trade-offs to his own advantage.

There are many individuals. Each individual is endowed with a unit of time which he allocates to leisure, l, and to labor, $n \equiv 1 - l$, which he sells in a competitive labor market. The productivity of labor varies among the individuals in the economy, so each does not receive the same wage. An individual with productivity x receives a wage of x and earns income, $y(x) = xn(x)$. The distribution of x, denoted $F(x)$, is fixed and known to all, with $F'(x) > 0$ for all x. There are people at all productivity levels; $F(x)$ is the fraction of the population

---

7. The relation to earlier literature was added in response to the useful comments of Thomas Borcherding. Lindbeck (1982) is an earlier effort at a positive analysis of the effects of spending and tax decisions on an individual's decision to work.

8. There is no money in our model. The cash-equivalent mix corresponds to the redistribution of general purchasing power, hence its designation. People who purchase both goods reach a welfare-maximizing mix.

with productivity less than x. The first of the two consumption goods is the numeraire, and wages and income are measured in terms of the numeraire.

Each good is produced using a separate constant returns-to-scale technology with (productivity adjusted) labor as the only factor of production. There is no capital and no uncertainty. Goods are offered in competitive markets at constant (before tax) prices 1 and p. Under constant returns to scale, these prices cannot vary with the supply of labor.

Each individual maximizes a strictly increasing and strictly concave utility function, $u(c_1, c_2, l)$, for the two consumption goods and leisure. The utility function is the same for all individuals. Both goods and leisure are normal goods with infinite marginal utility at the zero level of consumption and leisure. The individual's budget constraint is determined by his income, the taxes he pays, and the redistribution he receives.

Taxes are levied against the purchases of goods. The tax rates, t_1 and t_2, are a constant fraction of the expenditure on the respective good, so the after-tax market prices are $1 + t_1$ and $p(1 + t_2)$ per unit purchase of good one and good two, respectively. Tax rates must be chosen so that $1 + t_1 > 0$ and $1 + t_2 > 0$.

Tax revenues finance lump-sum redistribution of good one, r_1, and good two, r_2. There are no secondary markets, so goods cannot be resold. The publicly redistributed transfers r_1 and r_2 are nonnegative; r_1 and r_2 are the minimum levels of consumption, measured in units of goods, guaranteed to each person. Total consumption of goods is

$$c_1 = r_1 + a_1$$

and

$$c_2 = r_2 + a_2$$

where $a_1 \geq 0$ and $a_2 \geq 0$ are the amounts of good one and good two, respectively, purchased from income $y(x)$. Hence the individual's budget constraint is

$$y(x) = nx = a_1(1 + t_1) + a_2 p(1 + t_2) \tag{1}$$

Each individual acts as a pricetaker in the labor and goods markets, treats government policy choices as given, and chooses n, a_1, and a_2 to maximize utility. Formally, the problem for $x > 0$ is

$$\max_{a_1 \geq 0, a_2 \geq 0} u\left[r_1 + a_1, r_2 + a_2, 1 - \frac{(1 + t_1)a_1 + (1 + t_2)pa_2}{x}\right] \tag{2}$$

where we have solved (1) for n. The first-order conditions for (2) are

$$\frac{\partial u}{\partial a_1} = u_1 - \frac{(1 + t_1)}{x} u_3 \leq 0 \tag{3}$$

with equality for $a_1 > 0$ and

$$\frac{\partial u}{\partial a_2} = u_2 - \frac{(1+t_2)p}{x} u_3 \leq 0 \tag{4}$$

with equality for $a_2 > 0$.[9] The assumptions of a strictly concave utility function with a convex budget set ensure that the first-order conditions define a unique maximum. We denote the unique optimal choices as $a_1(x, z)$ and $a_2(x, z)$, where z is the vector of policy choices, $z = (r_1, r_2, t_1, t_2)$.

It will prove useful later to have the second-order conditions when $a_1 > 0$ and $a_2 > 0$. These conditions are that the matrix

$$H = \begin{bmatrix} h_{11} & h_{12} \\ h_{12} & h_{22} \end{bmatrix} \tag{5}$$

be negative definite, where

$$h_{11} = \frac{\partial^2 u}{\partial a_1^2} = u_{11} - 2\frac{(1+t_1)}{x} u_{13} + \frac{(1+t_1)^2}{x^2} u_{33} \tag{6}$$

$$h_{12} = \frac{\partial^2 u}{\partial a_1 \partial a_2} = u_{12} - \frac{p(1+t_2)}{x} u_{13} - \frac{(1+t_1)}{x} u_{23} + \frac{p(1+t_1)(1+t_2)}{x^2} u_{33} \tag{7}$$

and

$$h_{22} = \frac{\partial^2 u}{\partial a_2^2} = u_{22} - 2\frac{p(1+t_2)}{x} u_{23} + \frac{p^2(1+t_2)^2}{x^2} u_{33} \tag{8}$$

Of course, the strict concavity of u ensures that H is negative definite for any choices of c_1, c_2, and l, whether optimal or not.

People with low productivity choose not to work if redistribution is offered. From the first-order conditions (3) and (4) we find the critical value of x, x_0, that divides individuals into those who work and those who choose full-time leisure.

$$x_0(z) = \min\left[\frac{(1+t_1)u_3(r_1, r_2, 1)}{u_1(r_1, r_2, 1)}, \frac{p(1+t_2)u_3(r_1, r_2, 1)}{u_2(r_1, r_2, 1)}\right] \tag{9}$$

Those with $x \leq x_0(z)$ choose not to work and earn no income; $n(x) = 0$ for $x \leq x_0$. Those with productivity greater than x_0 work. For $x > x_0$, at least one good is purchased, so either (3) or (4) or both hold as an equality.

If redistribution is restricted to a single good, the number of people who choose full-time leisure is reduced to a minimum. In this case either $r_1 = 0$ or $r_2 = 0$; one of the marginal utilities, u_1 or u_2, becomes infinite, and the value of x_0 goes to zero. Everyone works who can work.

---

9. The assumption that $u_3(c_1, c_2, 0) = \infty$ ensures that $n < 1$ is a natural constraint.

In contrast, if redistribution is at the cash-equivalent mix, the number of people who choose full-time leisure is at a maximum. The reason is that each individual—including those who do not work—equates his marginal rate of substitution of good one for good two to the relative price ratio. Let r_1^* and r_2^* denote the cash equivalent transfer. Then,[10]

$$\frac{u_1(r_1^*, r_2^*, 1)}{u_2(r_1^*, r_2^*, 1)} = \frac{(1+t_1)}{p(1+t_2)} \qquad (10)$$

Obviously, at the cash-equivalent mix those who earn income and purchase one or both goods equate the marginal rate of substitution to the relative price ratio. At the cash-equivalent mix, people consume the mix of goods that maximizes utility, so

$$\frac{u_1}{u_2} = \frac{(1+t_1)}{p(1+t_2)} \quad \text{for all } x \qquad (11)$$

Given (11), the terms in the brackets in (9) are identical at the cash-equivalent mix, so x_0 specializes to

$$x_0^* = (1+t_1)\frac{u_3(r_1^*, r_2^*, 1)}{u_1(r_1^*, r_2^*, 1)} \qquad (12)$$

Everyone who works pays taxes on his purchases. The revenues collected by government—total tax collections—depend on tax rates, the labor-leisure choice, and the distribution of productivity, F. Recall that $F(x)$ is the fraction of the population with productivity less than x. Per capita government revenue, g, is given by

$$g = \int_{x_0}^{\infty} [t_1 a_1(x) + t_2 p a_2(x)] dF(x) \qquad (13)$$

---

10. This can easily be seen by solving the nonworkers' utility maximization problem when the redistributed goods can be resold:

$$\max u(c_1, c_2, 1)$$

subject to the budget constraint

$$c_1(1+t_1) + c_2 p(1+t_2) = r_1(1+t_1) + r_2 p(1+t_2)$$

The necessary and sufficient conditions for optimality are

$$\frac{u_1(c_1^*, c_2^*, 1)}{u_2(c_1^*, c_2^*, 1)} = \frac{(1+t_1)}{p(1+t_2)}$$

and the budget constraint. Clearly if $r_1 = r_1^*$ and $r_2 = r_2^*$, the nonworker will make no trades, even if he has the opportunity to do so.

where x_0 is given by (9). Since the budget is balanced, g finances per capita redistribution

$$g = r_1 + pr_2 \qquad (14)$$

Taken together, (13) and (14) are the government's budget constraint.

An equilibrium allocation must balance the government's budget at values for tax rates and redistribution consistent with individual utility maximization, equation (2). As is well known, the equilibrium allocation will not be Pareto optimal because we are not using lump-sum taxation. The allocation will be optimal, however, given the social contract represented by the voting rule for the set of linear taxes and lump-sum transfers that we consider.

There are many choices for equilibrium government policy within the social contract we have specified. For example, we have shown that restricting redistribution to a single good maximizes the number of people who choose to work. As the redistribution mix moves toward the cash-equivalent mix, more people choose their (unconstrained) optimal bundles. The number who choose to work decreases. We have not shown that it is rational to restrict choice and encourage participation in the labor force, however. To establish the condition under which restrictions of this kind are imposed, we turn to the choices made when people vote on the equilibrium government policy.

III. The Decisive Voter's Choice

The equilibrium government policy determines the relative size of government, measured by the share of national income spent or taken in taxes and the mix of goods redistributed. In this section we show that if the decisive voter chooses tax rates that are unequal, $t_1 \neq t_2$, then he will never choose the cash-equivalent redistribution, whatever his place in the distribution of voters. We defer until the next section the special case where $t_1 = t_2$ so that taxes are proportional to earned income.

Policies are chosen by means of a political process in which the decisive voter makes social choices. The political process yielding a decisive voter is modeled in the appendix. The method used is an extension of that used by Roberts (1977). Roberts shows that in a single good model, a voter's preference for the tax rate is inversely related to income if the ordering of individual income is independent of the choice of redistribution and taxes. Hence the median-income voter among the enfranchised citizens is decisive in a majority-rules election. In our extension to multiple goods we hold separate elections for the transfer-tax pairs (r_1, t_1) and (r_2, t_2). In the appendix we show that the median-productivity voter among the enfranchised citizens is decisive in each of these elections under rational expectations.[11]

11. Enelow and Hinich (1984, Chap. 8) prove that under uncertainty risk-averse voters who maximize expected utility can reach a voting equilibrium. Their result is conditional on expectations that are exogenous to the analysis.

The decisive voter chooses equilibrium government policy to maximize his own welfare. He is constrained in his choice by the requirements that the government's budget be balanced. Letting $x^d > 0$ be the productivity of the decisive voter, $a_1^d = a_1(x^d, z)$ and $a_2^d = a_2(x^d, z)$, the decisive voter's problem is to

$$\max_z u^d \equiv u\left[r_1 + a_1^d, r_2 + a_2^d, 1 - \frac{(1+t_1)a_1^d + (1+t_2)a_2^d}{x^d}\right] \quad (15)$$

subject to (13) and (14) and the requirement that $r_1 \geq 0$, and $r_2 \geq 0$.[12] Because of the inequality constraints and the complexity of the expression (13) for g, it is difficult to characterize the solution to (15). We are able to show, however, that for all x^d, the cash-equivalent redistribution mix is never the optimal choice when t_1 and t_2 differ.

Let ρ be the fraction of the government budget redistributed as good one and $(1-\rho)$ be the fraction redistributed as good two so that

$$r_1 = \rho g \quad \text{and} \quad r_2 = \frac{(1-\rho)g}{p} \quad (16)$$

Substituting (16) into (15), the decisive voter's problem becomes

$$\max_{t_1, t_2, \rho} u^d = u\left[\rho g + a_1^d, \frac{(1-\rho)}{p}g + a_2^d, 1 - \frac{(1+t_1)a_1^d + p(1+t_2)a_2^d}{x^d}\right] \quad (17)$$

where t_1, t_2, and $\rho \varepsilon [0,1]$ must be chosen so that $g \geq 0$.

At the cash-equivalent mix all people with productivity greater than x_0 work and buy both goods since, by assumption, both goods are normal and tastes are the same for all consumers. Hence both of the first-order conditions, (3) and (4), hold as equalities. To find the response of private spending to redistribution, we differentiate both (3) and (4) partially with respect to r_1, and then with respect to r_2. For $x > x_0^*$, we obtain

$$\begin{bmatrix} \partial a_1/\partial r_1 \\ \partial a_2/\partial r_1 \end{bmatrix} = H^{-1} \begin{bmatrix} -u_{11} + \frac{(1+t_1)}{x}u_{13} \\ -u_{12} + \frac{p(1+t_2)}{x}u_{13} \end{bmatrix} \quad (18)$$

---

12. If $x^d = 0$, this program reduces to

$$\max u^d = u(r_1, r_2, 1)$$

subject to (13) and (14).

and

$$\begin{bmatrix} \partial a_1/\partial r_2 \\ \partial a_2/\partial r_2 \end{bmatrix} = H^{-1} \begin{bmatrix} -u_{12} + \dfrac{(1+t_1)}{x} u_{23} \\ -u_{22} + \dfrac{p(1+t_2)}{x} u_{23} \end{bmatrix} \quad (19)$$

where H is given by (5).

Equations (18) and (19) show the effect of changes in r_1 and r_2 on the individual's consumption. The partial effect on government revenue, given in (13), of changes in the level of spending for r_1 or r_2 is (in vector notation),

$$I_1 \equiv \frac{\partial g}{\partial r_1} = \int_{x_0}^{\infty} (t_1 \, pt_2) \begin{pmatrix} \partial a_1/\partial r_1 \\ \partial a_2/\partial r_1 \end{pmatrix} dF(x) \quad (20)$$

and

$$I_2 \equiv \frac{1}{p} \frac{\partial g}{\partial r_2} = \frac{1}{p} \int_{x_0}^{\infty} (t_1 \, pt_2) \begin{pmatrix} \partial a_1/\partial r_2 \\ \partial a_2/\partial r_2 \end{pmatrix} dF(x) \quad (21)$$

Substituting (18) into (20) and (19) into (21), using equations (5) through (8) and algebra, we can calculate that at the cash equivalent mix,

$$(1+t_2)I_1 - (1+t_1)I_2 = -\int_{x_0^*}^{\infty} [t_1 \, pt_2] H^{-1} H \begin{bmatrix} 1+t_2 \\ -(1+t_1)/p \end{bmatrix} dF(x)$$

$$= (t_2 - t_1)[1 - F(x_0^*)] \quad (22)$$

To interpret this result, rewrite (22) as

$$\frac{I_1}{1+t_1} - \frac{I_2}{1+t_2} = \frac{\partial g}{\partial r_1(1+t_1)} - \frac{\partial g}{\partial r_2 p(1+t_2)}$$

$$= \frac{[t_2 - t_1][1 - F(x_0^*)]}{(1+t_1)(1+t_2)}$$

This equation shows the effect on government revenues of changes in redistribution evaluated at market prices. When $t_1 \neq t_2$, this effect can never be zero.

Now consider the total effect on government revenues of a change in ρ:

$$\frac{dg}{d\rho} = \frac{\partial g}{\partial r_1} \frac{dr_1}{d\rho} + \frac{\partial g}{\partial r_2} \frac{dr_2}{d\rho}$$

or

$$\frac{dg}{d\rho} = \frac{\partial g}{\partial r_1}\left(g + \rho \frac{dg}{d\rho}\right) + \frac{\partial g}{\partial r_2}\frac{1}{p}\left[-g + (1-\rho)\frac{dg}{d\rho}\right] \quad (23)$$

Collecting like terms in (23) we get

$$\frac{dg}{d\rho} A = g(I_1 - I_2) \tag{24}$$

where

$$A = \rho(1 - I_1) + (1 - \rho)(1 - I_2)$$

We can divide (24) by A to get [13]

$$\frac{dg}{d\rho} = \frac{g}{A}(I_1 - I_2) \quad \text{at } \rho = \rho^* \tag{25}$$

We now show that the cash-equivalent distribution, ρ^*, is never an optimal choice when $t_1 \neq t_2$. From (17) we calculate the decisive voter's marginal utility with respect to changes in ρ.

$$\frac{du^d}{d\rho} = u_1^d \left(g + \rho \frac{dg}{d\rho} \right) + u_2^d \frac{1}{p} \left(-g + (1 - \rho) \frac{dg}{d\rho} \right) \tag{26}$$

where we have used the first-order conditions (3) and (4) to eliminate terms containing $da_1^d/d\rho$ and $da_2^d/d\rho$. We set $\rho = \rho^*$ and use (11) to simplify (26).

$$\frac{du^d}{d\rho} = \frac{u_1^d}{1 + t_1} \left\{ g(t_1 - t_2) + [\rho^*(1 + t_1) + (1 - \rho^*)(1 + t_2)] \frac{dg}{d\rho} \right\} \tag{27}$$

Substituting (25) into (27) and then (22) into the result we get at $\rho = \rho^*$,

$$\frac{du^d}{d\rho} = \frac{u_1^d g}{(1 + t_1)A} [t_1 - t_2 + (1 + t_2)I_1 - (1 + t_1)I_2]$$

$$= \frac{u_1^d g}{(1 + t_1)A} (t_1 - t_2) F(x_0^*) \neq 0 \tag{28}$$

Hence $\rho = \rho^*$ is not a critical point for the decisive voter's utility maximization problem.

If the decisive voter has a unique optimal choice of ρ, we can compare his optimal policy to the cash-equivalent redistribution mix. When $t_1 > t_2$ we see from (28) that increasing ρ from ρ^* will improve the decisive voter's welfare. This is the normal case, where an increase in redistribution of either good by one dollar increases government revenue by less than one dollar.[14] With $t_1 > t_2$, the decisive voter chooses to transfer more of good one than in the cash-equivalent mix. In comparison with a cash-equivalent transfer, he takes more of his

---

13. We show by contradiction that A cannot be zero at $\rho = \rho^*$. Suppose that $A = 0$ so that by (24), $I_1 = I_2$. Then (22) reduces to $I_1 = 1 - F(x_0^*) < 1$, since $x_0^* > 0$, as seen from (12). This implies that $I_1 = I_2 < 1$, contradicting the supposition that $A = 0$.

14. $I_1 < 1$ and $I_2 < 1$, so $A > 0$ and $dg/d\rho > 0$, from (25).

consumption of good one as a transfer, and he taxes incremental purchases of good one at a high rate. Those above him in the income distribution pay the higher tax rate on all purchases of good one. The decisive voter taxes the second good at a lower rate or, by choosing $t_2 < 0$, subsidizes the consumption of that good. The decisive voter uses more of his income to purchase his consumption of good two, thereby avoiding the payment of the high taxes he imposes on purchases of good one.

IV. The Income Tax

One method of redistributing income, known as the negative income tax, has received considerable attention. The model can replicate principal features of the negative income tax. By setting $t_1 = t_2 = t$, the common tax rate on consumption goods becomes a tax on income. The cash-equivalent transfer, ρ^*, permits people to choose consumption bundles as if they received payment in cash. People with low productivity face the marginal tax rate (t) when they enter the labor force.[15]

The solution $t_1 = t_2 = t$ can arise as the decisive voter's optimal choice in solving (17) or as an exogenous constraint imposed to study income taxation. Under either interpretation, $du^d/d\rho = 0$ at $\rho = \rho^*$, as can be seen from (28). The issue, then, is whether ρ^* is a welfare minimum or maximum.

Under weak auxiliary assumptions we can show that ρ^* is a welfare-minimizing point if the decisive voter chooses to work. The additional assumption concerns the individual's labor supply behavior. We consider three cases: (1) an individual purchases both goods; (2) an individual consumes both goods but purchases only one of the goods; and (3) an individual does not work and consumes only what he or she receives as redistribution. We have already assumed that leisure is a normal good for all values of r_1, r_2, t, and x so that increased redistribution of income in any form reduces the labor supply. The precise effect of redistribution on the labor supply depends on whether the individual purchases both goods, a_1, $a_2 > 0$, or is in a corner with $a_1 = 0$ or $a_2 = 0$.

Consider, first, an individual who purchases both goods. Since he can substitute a_1 for a_2 in his private spending, neither consumption nor labor decisions are affected by incremental changes in the mix of goods redistributed. His optimal labor decision is, from (1),

$$n(x, z) = (1 + t)\frac{a_1 + pa_2}{x} \qquad (29)$$

Since leisure and both goods are normal goods, and the person buys both goods, any increase in redistribution permits him to achieve a higher level of consumption and leisure with less effort. Further, at the margin, the effect on his

---

15. Throughout, we neglect administrative costs. Those who favor a negative income tax usually claim lower administrative cost as an advantage.

decision to work is independent of the choice of the good redistributed.[16]

$$\frac{\partial n}{\partial r_1} = \frac{1}{p}\frac{\partial n}{\partial r_2} < 0 \tag{30}$$

The second case modifies the standard theory of the household to include an individual in a corner. The decisive voter's choice of redistribution leaves this voter in position where he does not purchase any of the good he receives as in-kind redistribution. He holds an excess supply of the good he receives. He would like to exchange some of the good he receives for the other good, but he is restricted from trading or exchanging.

For the case in which the recipient purchases good one but not good two, $a_1 > 0$, $a_2 = 0$. Then for given z,

$$\frac{\partial n}{\partial r_1} < \frac{1}{p}\frac{\partial n}{\partial r_2} \leq 0 \tag{31}$$

where

$$n = (1+t)\frac{a_1}{x} \tag{32}$$

There are now separate income effects for r_1 and r_2. The form redistribution takes matters to the recipient. If he receives more of the good he values least, r_2, the disincentive effect is smaller. Our assumption says that an increase of one dollar in either r_1 or r_2 causes the individual to work less, but the effect of increasing r_1 is a greater reduction of labor. The reason is that good one and good two are weak substitutes in consumption for individuals with $a_2 = 0$, since an increase in r_2 causes $c_2 = r_2$ to increase, but causes $c_1 = r_1 + a_1$ either to decrease or remain constant.[17]

16. Equation (30) follows from (18) and (19) with $t_1 = t_2 = t$.
17. Our results hold for a large class of utility functions. Substituting $a_2 = 0$ and (32) into the first-order condition (3) and differentiating gives

$$\frac{\partial n}{\partial r_1} = \frac{u_{11} - (1+t)x^{-1}u_{13}}{D}$$

and

$$\frac{p^{-1}\partial n}{\partial r_2} = \frac{u_{12} - (1+t)x^{-1}u_{23}}{pD}$$

where $D = x[-u_{11} + 2(1+t)x^{-1}u_{13} - (1+t)^2x^{-2}u_{33}]/(1+t)$, and D is positive by the concavity of u. Our assumption (31) requires that for $a_1 > a_2 = 0$ we have

$$u_{11} - (1+t)x^{-1}u_{13} < p^{-1}(u_{12} - (1+t)x^{-1}u_{23}) \leq 0$$

This condition is satisfied, for example, by any additively separable concave utility function. In the symmetric case, (33) requires that

$$p^{-1}(u_{22} - p(1+t)x^{-1}u_{23}) < u_{12} - p(1+t)x^{-1}u_{13} \leq 0$$

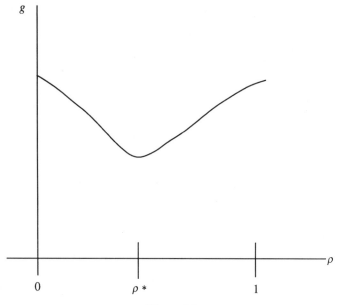

Figure 4-1.

The other case where $a_1 = 0$ and $a_2 > 0$ is completely symmetric, but with

$$\frac{1}{p}\frac{\partial n}{\partial r_2} < \frac{\partial n}{\partial r_1} \leq 0 \tag{33}$$

and

$$n = (1+t)\frac{pa_2}{x} \tag{34}$$

The effects just discussed are reversed.

In the third case, the recipient does not work. An increase in either r_1 or r_2 cannot induce him to enter the labor force.

Under the above assumptions on the labor supply, we show in the appendix that per capita government revenue reaches a minimum when the cash equivalent mix, ρ^*, is redistributed. The proof depends on the presence of workers who are in a corner. They purchase one good but not the other and, as in equations (31) and (33), they reduce labor less when they receive as redistribution the good that they do not purchase. Figure 4-1 shows this result. Per capita government revenue increases monotonically as we move away from ρ^*.

$$\frac{dg}{d\rho} \gtreqless 0 \quad \text{for} \quad p \gtreqless \rho^* \tag{35}$$

We have now shown why everyone is concerned about the form redistribution takes. The composition of redistribution affects incentives to work. Everyone receives his marginal product, but everyone pays taxes. Hence, for a given size or amount of redistribution, each taxpayer gains from increased participation in the labor force. The benefits are not distributed equally, however. The consumption of people who purchase both goods is unaffected by the form of redistribution, but these people benefit from increased participation in the labor force. At the opposite extreme are those with low productivity. They benefit most if redistribution is in cash, since they can consume their optimal bundles without working. The choice of ρ depends, therefore, on the productivity of the decisive voter.

We now show that if the decisive voter works, he will not choose a negative income tax, that is, the cash-equivalent redistribution mix. If the decisive voter does not work, his choice is ambiguous; we cannot determine whether he will choose a negative income tax—$\rho = \rho^*$—or redistribution in kind. We assume that the choices of ρ and t are unique.

To prove our proposition we return to the decisive voter's problem (17) with $t_1 = t_2 = t$. To have $g \geq 0$ we must have $t \geq 0$, and to ensure that $r_1 \geq 0$ and $r_2 \geq 0$ we must have $\rho \varepsilon [0, 1]$. With these constraints the decisive voter's first-order conditions are

$$\frac{du^d}{d\rho} = g\left(u_1^d - \frac{1}{p}u_2^d\right) + \left[\rho u_1^d + (1-\rho)\frac{1}{p}u_2^d\right]\frac{dg}{d\rho} \lesseqgtr 0 \quad \text{if } \begin{matrix} \rho = 0 \\ \rho \varepsilon (0,1) \\ \rho = 1 \end{matrix} \quad (36)$$

and

$$\frac{du^d}{dt} = \left[\rho u_1^d + \left(\frac{1-\rho}{p}\right)u_2^d\right]\frac{dg}{dt} - \frac{(a_1^d + pa_2^d)}{x^d}u_3^d \lesseqgtr 0 \quad \text{if } \begin{matrix} t = 0 \\ t > 1 \end{matrix} \quad (37)$$

We show by contradiction that if the decisive voter works he will not choose $\rho = \rho^*$. Suppose $\rho = \rho^*$. Since the decisive voter works, he purchases one or both of the goods at ρ^*. But at ρ^* we know the decisive voter is not at a corner so that $u_1^d = u_2^d/p$. Furthermore $u_1^d = u_2^d/p$ for an interval (ρ_1, ρ_2) that contains ρ^*. This is so because the voter can substitute privately for changes in the mix of goods publicly provided. From (36), for fixed t, utility is decreasing as ρ moves toward ρ^* when $u_1^d = u_2^d/p$. Hence ρ^* cannot be utility maximizing. The decisive voter chooses $\rho \neq \rho^*$.

If the decisive voter purchases both goods, he will redistribute only one good. Since he chooses to purchase both goods we know from (11) that $u_1^d = u_2^d/p$, so (36) reduces to

$$\frac{du^d}{d\rho} = u_1^d \frac{dg}{d\rho} \quad (38)$$

From (35) and (38) we see that if the decisive voter chooses to purchase both goods he chooses either $\rho = 0$ or $\rho = 1$, whichever gives the highest government revenues.

If the decisive voter does not work, we cannot determine whether he will choose a negative income tax or redistribution in kind. The reason for this ambiguity can be seen from equation (36). If $\rho < \rho^*$, the nonworker is receiving too much of good two relative to good one to be at a private optimum, that is, $u_1^d - (u_2^d/p) > 0$. Hence the first term on the right-hand side of equation (36) is positive. The second term is negative, since $dg/d\rho < 0$ for $\rho < \rho^*$. A symmetric argument holds for $\rho > \rho^*$.

The ambiguity arises from these two opposing effects. If the nonworker considers only his private choice, he equates u_1^d to u_2^d/p and chooses the cash-equivalent mix. Recognizing the general equilibrium effect of the budget can make him change his decision. By choosing (some) redistribution in kind, he can induce people to work, raising g and increasing the amount redistributed. We denote by ρ the nonworker's choice of ρ.

The optimal tax rate for a nonworker is unambiguous, however. Since $a_1 = a_2 = 0$, (37) requires that $dg/dt = 0$ at the optimal tax rate for a nonworker. A decisive nonworker sets taxes to maximize government revenue, since he pays no tax. Given ρ, this maximizes the total value of redistribution received.

In contrast, a decisive voter who chooses to work will not maximize government revenue, since either $a_1^d > 0$ or $a_2^d > 0$, or both. In the case where the decisive voter buys both goods, substituting (1), (3), and (4) into (37) we get

$$(1 + t)^2 \frac{dg}{dt} = y^d \qquad (39)$$

where y^d is the decisive voter's optimal income. It is revealing to rewrite (39) in terms of average income, $\bar{y} = [(1 + t)/t]g$:

$$\bar{y} + t(1 + t)\frac{d\bar{y}}{dt} = y^d \qquad (40)$$

Let $\bar{x}$ denote the productivity of an individual earning the average income, $\bar{y}$. From (40) we see that if $x^d \geq \bar{x}$, the decisive voter neither taxes nor redistributes; $t = 0$ is the optimal choice. If $x^d < \bar{x}$, then $t > 0$ is optimal.[18]

Figure 4-2 summarizes these decisions and shows the relation between the productivity of decisive voters and their choice of tax rate and type of redistribution. If the voting rule restricts voting to people with above-average incomes, $x^d > \bar{x}$, there is no redistribution.[19]

Below $\bar{x}$ the decisive voter receives net benefits from redistribution. He votes for taxes and redistribution, but he purchases both goods, so he favors maximum participation in the labor force. To maximize his utility he sets the tax

---

18. This result replicates the conclusion in Meltzer and Richard (1981). It requires, as before, that y is nondecreasing in x for all z.

19. The linear increasing and decreasing segments in the upper and lower panels of Figure 4-2 are a convenience. There is no presumption that the relation is linear.

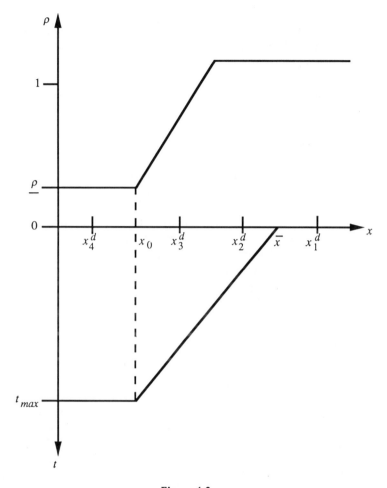

Figure 4-2.

rate low and redistributes one good; either $\rho = 0$ or, as in Figure 4-2, $\rho = 1$. If his productivity is somewhat lower but he still works, the decisive voter redistributes both goods but not in the cash-equivalent mix. Finally, if the decisive voter does not work, $x^d \leq x_0$, he chooses the tax rate that maximizes redistribution and chooses $\rho = \bar{\rho}$. The four representative positions are shown in Figure 4-2 as $x_1^d, \ldots, x_4^d$.

V. Conclusion

Normative arguments for restricting redistribution to cash transfers or a negative income tax have not appealed to politicians or their constituents. All modern governments redistribute income in kind and provide goods and services that can be produced and distributed privately. At times public and private supply coexist. Housing services, medical care, and education are examples, but so too are safety and many regulatory activities of government.

(Meltzer & Richard, 1981) identified the size of government with the share of government spending in total output and showed the conditions under which a decisive voter determines the size of government by choosing the tax rate and amount of income redistribution that maximize his utility. The choice of a decisive voter depends on the voting rule that a society adopts. In a society operating under strict majority rule, with linear taxes and lump-sum redistribution, the median voter dominates the choice of tax rate and redistribution.

In our earlier work and in much of the literature on the theory of redistribution, it is optimal to redistribute general purchasing power or the composite consumption good. Yet, as just noted, redistribution never consists solely of cash transfers. This chapter extends our study of optimal choice by decisive voters to consider their choice of redistribution mix and tax mix. Tastes are identical. Taxes are proportional to consumption. The decisive voter sets the tax rates to balance the government budget and maximize his utility. There are no public goods; all government spending is for redistribution by lump-sum transfers. No one is permitted to resell goods.

When choosing the type of redistribution he prefers, the decisive voter reasons as follows: If I allow each recipient to choose the bundle of goods and services that maximize his utility, those below some level of income will decide not to work. By giving income in kind, I can encourage people who are net beneficiaries of the tax-transfer system to make different labor-leisure choices. For example, if I allow the government to distribute only one good, people with relatively low productivity will find that all of their disposable income is in one consumption good; they are surfeited with the good that the government offers and have none of the other good(s). Since they cannot resell the good they receive, they will be willing to trade leisure for income to purchase the good(s) they do not have. By working they trade leisure, valued at the marginal product of labor, for consumption of a good that has relatively high marginal utility. They will continue to trade leisure for consumption of the good they purchase until they reach a constrained optimum. At the constrained optimum, they prefer to sell some of the good they receive as a transfer and to increase leisure, but they cannot. They may grumble about the welfare system, but they will prefer the constrained outcome to any other feasible point. Since they work and produce more goods and services and pay taxes, everyone else, including me, is better off. It is in my interest to get them to work more than they would choose to work if I permit redistribution of cash or a cash equivalent that provides their preferred consumption bundle.

By reasoning in this way, decisive voters reach a more general conclusion about redistribution. The welfare system can increase the amount of work, and the amount of earned income and tax receipts at given tax rates, by making transfers in kind. If both goods are normal, people will work to increase their consumption of the good for which marginal utility is greater than price. In fact, if decisive voters choose to tax the consumption of goods at different rates, they will never redistribute cash or generalized purchasing power.

More insights are available in the special case where the decisive voter chooses (or is constrained to choose) an income tax.[20] He knows that everyone maximizes, subject to constraints. If he raises the tax rate, after-tax wages fall. Everyone earns less, but more income is available for redistribution. He continues to raise taxes until the loss from the disincentive effect of higher tax rates exceeds the amount he gains from additional redistribution and leisure. If he raises the tax rate beyond his optimum, he and all those who work choose less labor and more leisure. The additional consumption obtained from his lower earned income and the higher level of redistribution is worth less to him than the gain from additional leisure. Conversely, if he lowers the tax rate he moves to a position that he can improve by working less and receiving a larger transfer.

Obviously, the position of the decisive voter depends on the voting rule. A two-thirds majority or a property qualification for voting lowers the equilibrium tax rate. Extensions of the franchise that permit more low-income recipients to vote raise the tax rate and increase the number of voters favoring redistribution of cash. Shifts in the income distribution that raise the decisive voter's income closer to the mean of the income distribution reduce taxes and spending and increase the proportion of redistribution that is offered as a particular good. Further, transfers to individuals that increase the number of nonworkers also increase votes for higher taxes and against in-kind transfers.[21] Our analysis suggests why most welfare payments in the United States are in-kind transfers. The principal exceptions are cash payments to those unlikely to enter the labor force under any circumstances, such as retired and disabled workers.

Under an income tax, for given tastes, a given distribution of productivity, and a voting rule, the decisive voter chooses one of four outcomes:

1. If the decisive voter does not work at the chosen tax rate, he is part of a majority with identical tastes who prefer to consume the amount redistributed and enjoy full-time leisure. The decisive voter chooses the tax rate that maximizes the amount available for redistribution. Only in this case is a negative income tax possible.

2. At the opposite extreme, the franchise is restricted. The decisive voter has above-average income, so he votes no taxes to pay for redistribution. There is no redistribution.

3. If the decisive voter works and consumes more of each good than the maximum he can receive under any feasible transfer, the composition of his consumption is unaffected by the composition of redistribution. He is aware, however, that his choices change the incentives others face. The decisive voter distributes only one of the two goods to ensure that the marginal utility of the remaining good is substantially larger than the marginal disutility of labor. His decision to redistribute in kind increases participation in the labor force by forcing people with low productivity into a corner solution. If only one good is redistributed, all ablebodied people work.

20. The choice of different tax rates for different goods is an argument for indirect rather than direct taxes. The decisive voter chooses tax rates as part of his individual welfare maximizing choice of the amount and type of redistribution.

21. Evidence of the effect of changes in the ratio of mean income to the income of the decisive voter and in the proportion of the population receiving welfare is in Meltzer and Richard (1983).

4. If the distribution of either good exceeds the decisive voter's desired consumption of that good, the decisive voter is in a corner. It is in his interest to distribute both goods. As the distribution of the second good increases, he moves toward his individual optimum, but by redistributing both goods, he lowers the marginal utility of the second good and reduces incentives to work. The choice he makes reflects the loss he bears as the aggregate amount of labor declines and the benefit he obtains by moving toward a position at which the ratio of the (private) marginal utilities in consumption equals the ratio of the market prices.

The analysis shows why the equilibrium reached in a political-economic system differs from the equilibrium reached in a market economy and in discussions of negative income taxation. The decisive voter does not choose redistribution to maximize the welfare of the recipients; he maximizes his own utility. His choice reflects the constraints imposed by the political-economic environment. These include the effects on everyone's incentives and decisions to work and pay taxes. Because the budget is balanced in real terms and the identity of the decisive voter is independent of the number of people who work, the decisive voter benefits from the induced increase in aggregate income.

Appendix

The Voting Equilibrium

The extension of Roberts' (1977) results to choosing r_1, r_2, t_1, and t_2 is straightforward under rational expectations. We consider two separate elections: the first for pairs (r_1, t_1), holding fixed (r_2, t_2), and the second for pairs (r_2, t_2), given the outcome of the first election. Suppose all voters anticipate that the outcome of the second election will be (r_2^*, t_2^*). Then totally differentiating the utility function (2) with respect to r_1 and t_1 and using the first-order condition (3) and (4) to simplify, we get

$$du = u_1 \left(dr_1 - \frac{a_1}{1 + t_1} dt_1 \right)$$

This equation is equivalent to Roberts' equation (6). Since both goods are normal, a_1 is nondecreasing in x for all choices of r_1, r_2, t_1, and t_2. Hence Roberts' hierarchical adherence assumption is satisfied, implying that the social choice of (r_1, t_1) is ordered by productivity x. Therefore the voter with median productivity among the enfranchised citizens is decisive in a majority-rule election for (r_1, t_1).

Denote the decisive voter's choices by $r_1^* = r_1^*(r_2^*, t_2^*)$ and $t_1^* = t_1^*(r_2^*, t_2^*)$. Now given (r_1^*, t_1^*) we can repeat the above analysis, mutatis mutandis, to conclude that the same median voter among the enfranchised citizens is decisive in a majority-rule election for (r_2, t_2). Rational expectations require that (r_2^*, t_2^*) be the decisive voter's choice. We conclude that there is a rational-expectations equilibrium in which the social choices are the decisive voter's personally optimal choices.

There are two unique features of our model which generate the voting equilibrium. First, people differ only along one dimension, namely, their productivity. Second, we hold separate elections for each transfer-tax pair, tax (r_i, t_i), $i = 1, 2$. It should be clear that our results can be extended easily to separate elections for any number of transfer-tax pairs, (r_i, t_i), $i = 1, 2, \ldots, n$. In this extension the decisive voter's choices continue to be an equilibrium.

Proof that Government Revenue Is Minimized at the Cash-Equivalent Mix

We now show that per capita government revenue, g, is a minimum for $\rho = \rho^*$. That is, $dg/d\rho > 0$ for $\rho > \rho^*$ and $dg/d\rho < 0$ for $\rho < \rho^*$.

If $\rho > \rho^*$, the government is supplying a mix of redistribution that contains more good one than some workers desire given their wage. One constraint is binding: there is a set of workers who purchase good two, but not good one. Let $x_1(\rho, t) > x_0$ be the productivity of the last worker not to purchase good one. Hence, for $x \varepsilon (x_0, x_1)$, $a_2 > 0$, $a_1 = 0$, and $n(x) = (1 + t)pa_2/x$. For $x > x_1$, $a_1 > 0$, $a_2 > 0$, and $n(x) = (1 + t)(a_1 + p a_2)/x$. Per capita government revenue is

$$g = \int_{x_0}^{x_1} t p a_2(x) dF(x) + \int_{x_1}^{\infty} t[a_1(x) + p a_2(x)] dF(x) \tag{A1}$$

Substituting (29) and (34) into (A1) we get

$$g = \frac{t}{1+t} \int_{x_0}^{\infty} xn \, dF(x) \tag{A2}$$

Therefore

$$\frac{dg}{d\rho} = \frac{t}{1+t} \int_{x_0}^{\infty} x \frac{dn}{d\rho} dF(x) \tag{A3}$$

Now

$$\frac{dn}{d\rho} = \frac{\partial n}{\partial r_1}\left(g + \rho \frac{dg}{d\rho}\right) + \frac{1}{p}\frac{\partial n}{\partial r_2}\left[-g + (1-\rho)\frac{dg}{d\rho}\right]$$

and using (29) this reduces to

$$\frac{dn}{d\rho} = \begin{cases} g\left(\frac{\partial n}{\partial r_1} - \frac{1}{p}\frac{\partial n}{\partial r_2}\right) - \frac{dg}{d\rho}\left[(1-\rho)\left(\frac{\partial n}{\partial r_1} - \frac{1}{p}\frac{\partial n}{\partial r_2}\right) - \frac{\partial n}{\partial r_1}\right] & \text{for } x_1 > x > x_0 \\ \\ \frac{dg}{d\rho}\frac{\partial n}{\partial r_1} < 0 \quad \text{for } x \geqslant x_1 \end{cases} \tag{A4}$$
$$\tag{A5}$$

By (33) both terms in brackets in (A4) are positive. Substituting (A4) and (A5) into (A3) gives

$$\frac{dg}{d\rho} = \frac{\dfrac{tg}{1+t}\int_{x_0}^{x_1} x\left[\dfrac{\partial n}{\partial r_1} - \dfrac{1}{p}\dfrac{\partial n}{\partial r_2}\right]dF(x)}{1 + \dfrac{t}{1+t}\int_{x_0}^{x_1} x(1-\rho)\left[\dfrac{\partial n}{\partial r_1} - \dfrac{1}{p}\dfrac{\partial n}{\partial r_2}\right]dF(x) - \dfrac{t}{1+t}\int_{x_0}^{\infty} x\dfrac{\partial n}{\partial r_1}dF(x)} > 0$$

Therefore $dg/d\rho > 0$ for $\rho > \rho^*$.

By reversing the roles of good one and good two in the above proof, we can show that $dg/d\rho < 0$ for $\rho < \rho^*$. Furthermore, when $\rho = \rho^*$, no workers are in the corner so $x_1 = x_0$ and $dg/d\rho = 0$ at $\rho = \rho^*$. Therefore we have shown that the cash-equivalent redistribution mix ($\rho = \rho^*$) resultx in the minimum per capita government revenue for any given t.

References

Anderson, M. (1978). *Welfare*. Stanford, Conn.: Hoover Institution.
Atkinson, A. B. and Stiglitz, J. E. (1980). *Lectures on Public Economics*. New York: McGraw-Hill.
Corlett, W. J. and Hague, D. C. (1953). "Complementarity and the Excess Burden of Taxation," *Review of Economic Studies*, 21, pp. 21–30.
Diamond, P. A. and Mirrlees, J. A. (1971). "Optimal Taxation and Public Production." *American Economic Review*, 61, pp. 8–27, 261–78.
Enelow, J. M. and Hinich, M. J. (1984). *The Spatial Theory of Voting*. Cambridge: Cambridge University Press.
Friedman, M. (1962). *Capitalism and Freedom*. Chicago: University of Chicago Press.
Friedman, M. and Friedman, R. (1980). *Free to Choose*. New York: Harcourt, Brace, Jovanovich.
Harberger, A. C. (1964). "Taxation, Resource Allocation and Welfare." *The Role of Direct and Indirect Taxes in the Federal Revenue System*. Princeton, N.J.: Princeton University Press (for the National Bureau of Economic Research and the Brookings Institution).
Hochman, H. and Rodgers, J. D. (1969). "Pareto Optimal Redistribution." *American Economic Review*, 59 (4), pp. 542–57.
Lindbeck, A. (1982). "Tax Effects versus Budget Effect on Labor Supply." *Economic Inquiry*, 20, pp. 473–89.
Little, I. M. D. (1951). "Direct versus Indirect Taxes." *Economic Journal*, 56, pp. 38–50.
Meltzer, A. H. and Richard, S. F. (1981). "A Rational Theory of the Size of Government." *Journal of Political Economy*, 89, pp. 914–27.
Meltzer, A. H. and Richard, S. F. (1983). "Tests of a Rational Theory of the Size of Government." *Public Choice*, 41, pp. 403–18.
Pommerehne, W. W. (1979). "Gebundene vs. freie Geldtransfers: Eine Fallstrudre. C. C. von Weizsacker (ed.), *Staat und Wirtschaft*. Berlin: Duncker and Humblot.
Ramsey, F. P. (1927). "A Contribution to the Theory of Taxation." *Economic Journal*, 37, pp. 47–61.

Roberts, K. W. S. (1977). "Voting over Income Tax Schedules." *Journal of Public Economics*, *8*, pp. 329–40.
Romer, T. (1975). "Individual Welfare, Majority Voting and the Properties of a Linear Income Tax." *Journal of Public Economics*, *4*, pp. 163–85.
Tobin, J. (1966). The Case for an Income Guarantee." *The Public Interest*, *4*.
Tobin, J., Pechman, Joseph, A., and Mieszkowski, Peter M. (1967). "Is a Negative Income Tax Practical?" *Yale Law Journal*, *77*, pp. 1–27.

5

A Political Theory of Progressive Income Taxation

ALEX CUKIERMAN AND ALLAN H. MELTZER

I. Introduction

Progressive income taxation is found in all developed and in many developing countries. In most of these countries both average and marginal income tax rates increase with the level of income. Yet this ubiquitous phenomenon has proved troublesome for economists and social scientists. Despite numerous attempts to make the case for or against progressivity, and many strong statements on both sides, the rational case for progressivity has proved elusive.

Over a generation ago, Blum and Kalven (1953) pointed out that most arguments for progressivity have a weak foundation. Many rely on comparisons of marginal utility of income across individuals, an assumption that economists reject along with other interpersonal comparisons of utility. After considering the arguments, Blum and Kalven (p. 71) concluded that the strongest case for progressive taxes depends on the argument that progressivity provides revenues for redistribution through the government budget. There is now broad agreement that the case for redistribution and tax progressivity cannot be made on strictly economic, nonpolitical grounds.[1]

Despite this broad agreement, much recent research on tax progressivity focuses on the conditions that would lead to the choice of progressive taxes as an optimal form of taxation. This work is normative, not positive, and much of it is based on the utilitarian principle of maximizing the sum of individual utilities.

We benefited from useful discussions with Abhijit Banerjee, Soren Blomquist, Avinash Dixit, Tom Romer, Howard Rosenthal, and Ephraim Sadka and from the criticisms of several referees on a previous version. Parts of the previous version were written while Meltzer was a visiting scholar at the Bank of Japan. Their support is gratefully acknowledged.

1. See the comments by James Tobin, Allen Wallis, Oswald Brownlee, Norman Ture, and Richard Musgrave in Campbell (1977).

Pigou (1947) showed that this principle leads to extreme progressivity; his optimal tax policy is full equalization of after-tax incomes. Mirrlees (1971) showed that Pigou's conclusion changes considerably when there are incentive effects of taxation on the choice between labor and leisure. Tax rates are much lower, and tax schedules are either linear or rates fall as income rises (Mirrlees, 1971; Phelps, 1973; Sadka, 1976).[2] Atkinson (1973) imposed a social utility function: the government chooses to alter the distribution of income by lowering inequality, as Simons (1938, pp. 18–19) had urged. Progressivity can be obtained in this case if the government is willing to move the economy to a Pareto inferior position.

The optimal tax literature typically imposes a utility function and derives the tax function. An alternative approach taken by Romer (1975) and Roberts (1977) is to specify a tax function and allow the voters to choose the parameters of this function through majority rule. Romer (1975) showed that, if individuals differ in ability, the decisive voter is the individual with median ability. The decisive voter chooses increasing average progressivity (Romer, 1975), if the utility function is Cobb–Douglas. Meltzer and Richard (1981) extended this result for net redistribution to a large class of utility functions and provided evidence (Meltzer & Richard, 1983) that the model is broadly consistent with U.S. data.

Tax schedules and so-called effective tax rates typically rise with the level of income. Marginal tax rates often rise also. Table 5-1 compares the effective tax rate paid by a family with two children that earns the mean level of income to the tax rates paid by a comparable family that earns two or four times mean income. In most countries, the average effective tax rate increases as income rises.[3]

Table 5-1 also shows that countries have very different tax functions. Denmark, Germany, and Sweden have similar per capita income and all three have relatively high average tax rates at mean income. Progressivity differs, however; it is greater in Sweden and Denmark than in Germany. Australia, Japan, and Canada have relatively low tax rates and, again, marked differences in progressivity. The Australian increase in tax rates with income is similar to Sweden's, but the average rates are much lower.

Differences in tax rates are often associated with differences in spending for redistribution. Meltzer and Richard (1981), using a linear tax function, showed that the decisive voter's choice of per capita transfer payments determines the tax rate. Net tax payments are negative at low incomes and for nonworkers, and rates rise with the level of income in their analysis.

---

2. An early survey of this literature is in Atkinson (1973). Mirrlees' (1971, p. 186) own summary is: "The optimum tax schedule depends upon the distribution of skills in such a complicated way that it is not possible to say in general whether marginal rates should be higher for high income, low income or intermediate income groups."

Linearity of the optimal schedule extends to the case in which individuals respond to higher tax rates by working in an untaxed sector of the economy (Kramer & Snyder, 1983, 1984).

3. We believe this is true in the United States also, although the table is restricted to countries for which the Organization for Economic Cooperation and Development (OECD) attempts to provide comparable data.

Table 5-1. Average tax rates of a family with two children filing a joint tax return at various income levels, 1974[a]

Country	100%	Normalized Gross Income Level 200%	400%
Australia	7.4	20.1	35.7
Austria	13.2	18.6	24.1
Belgium	15.2	22.5	30.4
Canada	9.8	19.5	28.4
Denmark	31.0	43.3	52.7
Finland	22.9	32.5	43.5
France	8.4	12.8	16.9
Germany	22.9	29.5	35.4
Ireland	15.7	24.1	36.5
Italy	7.7	12.4	23.7
Japan	8.4	12.1	18.2
New Zealand	15.6	24.7	36.0
Norway	21.2	31.3	44.0
Spain	7.0	8.4	15.0
Sweden	24.4	36.2	50.5
Switzerland	17.3	25.9	34.6
United Kingdom	13.7	26.0	30.9

[a]Gross income is expressed as a percentage of an average production worker's earnings. The tax rates are for a family income that is contributed in equal shares by both spouses and they include both personal income taxes and social security contributions. Data are available in the source for other earning profiles.

Source: Committee on Fiscal Affairs, Organization for Economic Cooperation and Development, (1978). *The Tax/Benefit Position of Selected Income Groups in OECD Member Countries 1972—1976.* Paris: OECD, Table 16(b), p. 110.

This chapter provides a positive theory of progressivity for marginal and average tax rates within a majority-rule framework by using a tax function that permits marginal progressivity, linearity, or regressivity. The analysis suggests some reasons for observed differences in marginal and average tax rates.

The analysis develops a set of conditions under which majority voting implies marginal progressivity. We assume that each person has the same utility function, but that people differ in ability and therefore in productivity. Under majority rule, the choice of progressivity depends, partially, on the response of labor supply to tax rates. If people with high ability show small response of labor supply to tax rates, majority rule is likely to produce marginal progressivity. Marginal progressivity is not restricted to this case, however. Even when the response of work effort to an increase in tax burden is not systematically related to income levels, the decisive voter can choose marginal progressivity. He is more likely to do so the higher the variance of gross incomes and the more skewed to the right the distribution of gross incomes. Studies of the distribution of income show that the distribution is indeed skewed to the right. The variance and the degree of positive skewness in the distribution of gross incomes are in turn larger the larger the variance and the degree of positive skewness in the distribution of abilities.

To allow the majority-rule process to pick a (possibly) progressive tax schedule it is necessary to extend the family of linear tax schedules used in previous literature to a three-parameter family. The government's budget constraint requires that the budget be balanced. This constraint determines one of these parameters as a function of the other two, and majority rule picks the remaining two parameters out of the set of feasible pairs of such parameters. It is well known that majority voting over a multidimensional issue space of this kind may induce collective intransitivities that preclude the existence of a stable, unique majority winner. This is an aspect of Arrow's (1951) impossibility theorem. Hence it is necessary to determine whether a majority-winning tax schedule exists. This task logically precedes that of finding conditions for progressivity since, in the absence of a majority winner, majority rule does not produce a well-defined choice of tax schedule.

We show that if the ranking of incomes is independent of tax schedules (which is the case when both consumption and leisure are normal goods), the set of local majority-winning schedules is not empty. Moreover this set contains the set of feasible schedules most preferred by the individual who is at the median of the distribution of abilities. Hence the individual with median ability is locally decisive. This individual is also globally decisive for a wide range of utility functions, provided the proportion of individuals with intermediate levels of ability is sufficiently large.

Section II analyzes the labor-leisure choice of individuals with different abilities who face a tax schedule with a given degree of progressivity or regressivity. Section III introduces the government budget constraint and shows that, when the ranking of gross incomes is independent of the tax schedule, the individual with median ability is locally decisive for the choice of tax schedule. Conditions under which this individual is also globally decisive over all feasible tax schedules are discussed in Section IV. Section V characterizes the choice of tax schedule by the person with median ability who works and also provides conditions under which he will pick a progressive tax schedule. Section VI briefly considers the same issues when the decisive voter does not work. Section VII shows, by means of an example, that conditions that are likely to produce progressivity when the budget is used for redistribution of income are also likely to lead to progressivity when the budget is used to finance a public good. A conclusion summarizes main results.

II. The Private Economy

The economy consists of a large number of individuals who differ in ability and therefore in their real wage rate. Each individual takes his wage rate and the tax schedule as given and chooses the amount of leisure, work, and consumption to maximize utility. Utility of the representative individual is given by a strictly concave function $u(c, l)$ of consumption c and leisure l. Consumption is a normal good, and the marginal utility of consumption or leisure is infinite when the level of either consumption or leisure is zero.

Individual incomes reflect differences in individual productivity and the use of a common, constant returns-to-scale technology to produce the consumption good. An individual with productivity x earns pretax income y;

$$y(x) = xn(x) \tag{1}$$

where $n(x)$ is the amount of work he supplies. Each individual is endowed with one unit of time that he can allocate either to leisure, $l(x)$, or to the production of the consumption good, so $l(x) = 1 - n(x)$.

Tax revenues finance a fixed level of government expenditures G. The total tax paid by an individual with gross income y is

$$T(y) = -r + \tau y + \alpha y^2 \tag{2}$$

where r, τ, and α are parameters of the tax schedule that are determined by the political process. The corresponding marginal tax schedule is

$$T'(y) = \tau + 2\alpha y \tag{3}$$

The parameter α measures the degree of marginal progressivity of the tax schedule. The marginal tax rate increases with income when α is positive and decreases with income when α is negative.[4] For $\alpha = 0$, equation (2) reduces to the widely used linear income tax schedule (Roberts, 1977; Romer, 1975; Sheshinski, 1972), and the marginal rax rate is constant at τ. A positive r corresponds to the case in which low-income people get a subsidy by means of a negative income tax or cash transfer. This is the case discussed by Meltzer and Richard (1981). When $\alpha = 0$ and $r > 0$, our model reduces to theirs.

We restrict the tax schedule in three ways. First, negative marginal tax rates are excluded. Second, marginal tax rates are smaller than 100 percent. Third, no individual pays more than 100 percent in taxes, so $r \geq 0$. Equation (4) summarizes the first two restrictions. The restrictions

$$0 \leq T'(y) = \tau + 2\alpha y < 1 \tag{4}$$

implicitly impose upper and lower bounds on the degree of marginal progressivity ($\alpha > 0$) and marginal regressivity ($\alpha < 0$), respectively.[5]

---

4. A sufficient condition for both marginal and average progressivity is $\alpha > 0$ and $r \geq 0$.

5. Let $x_u = \max x$ be the highest ability level in the population. Since each individual supplies at most one unit of labor, the maximum income of any individual is x_u. A sufficient condition for the inequalities in equation (4) is:

$$0 \leq \tau + 2\alpha x_u \leq 1 \quad \text{for all } y \leq x_u$$

This expression imposes the following bounds on α

$$\frac{-\tau}{2x_u} \leq \alpha \leq \frac{1-\tau}{2x_u}$$

There is no saving; consumption equals disposable income as shown in (5).

$$c(x) = r + (1 - \tau)xn - \alpha(xn)^2 \tag{5}$$

Given the wage rate, x, and the tax parameters r, τ, and α, individuals choose the allocation of their time and their consumption by solving

$$\max_n u[r + (1 - \tau)xn - \alpha(xn)^2, 1 - n] \tag{6}$$

When $r = 0$, there is no redistribution. Everyone works because we have assumed that the marginal utility of consumption is infinite when consumption is zero. Since the the marginal utility of leisure is also infinite at zero leisure, the solution to the problem in (6) is an internal one for all x. The first-order condition is

$$x[1 - T'(y)]u_c[c(x), 1 - n(x)] - u_l[c(x), 1 - n(x)] = 0 \tag{7}$$

where

$$1 - T'(y) \equiv 1 - \tau - 2\alpha xn \tag{8}$$

and $c(x)$ is given by (5). Equation (7) determines n as a function of individual productivity x and the tax parameters r, τ, and α.[6]

Individuals with productivity below a minimum level, denoted x_0, do not work. Their earned income is zero, and their consumption is r. The value of x_0, which divides the population into workers and nonworkers, is found from (7) to be

$$x_0 = \frac{u_l(r, 1)}{(1 - \tau)u_c(r, 1)} \tag{9}$$

The value of x_0 depends on r and τ but not on α since, at x_0, the person chooses full-time leisure and $xn = 0$. The effects of r and τ on x_0 are given by (10) and (11).

$$\frac{\partial x_0}{\partial r} = \frac{u_{lc}(r, 1) - (1 - \tau)x_0 u_{cc}(r, 1)}{(1 - \tau)u_c(r, 1)} \tag{10}$$

$$\frac{\partial x_0}{\partial \tau} = \frac{x_0}{1 - \tau} \tag{11}$$

When the redistribution parameter is zero, everyone works; the "last" worker is the individual with lowest ability in the population.

---

6. The second-order condition for a maximum is $D \equiv x^2 b^2 u_{cc} - 2xbu_{cl} + u_{ll} - 2\alpha x^2 u_c < 0$, where $b \equiv 1 - \tau - 2\alpha xn = 1 - T'(y)$.

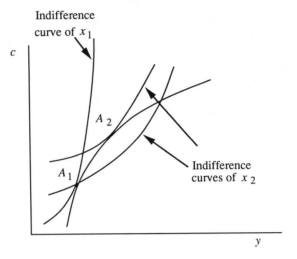

Figure 5-1.

To characterize the political equilibrium of the next section it is necessary to find conditions under which individuals with higher ability also have higher gross incomes or, to be more precise, the conditions under which the ordering of productivity corresponds to the ordering of earned income. The relation of the two orderings depends on the properties of the utility function.

It can be shown that the two orderings correspond if consumption and leisure are normal goods. For $\alpha \geqslant 0$ this is consequence of lemma 1 in Sadka (1976). Sadka shows that under the specified conditions the indifference curve (in the (c, y) plane) of a higher ability individual is less steep than that of a lower ability individual at *every* combination of consumption and of gross income.[7] His Figure 5-1 is reproduced (in our notation) as Figure 5-1 here. In the (c, y) plane all individuals face the same budget line which, from (5), is

$$c = r + (1 - \tau)y - \alpha y^2 \equiv y - T(y) \qquad (12)$$

The restriction in (4) in conjunction with $\alpha \geqslant 0$ implies that the budget line has a positive and nonincreasing slope as drawn in Figure 5-1. Consider now two individuals with productivity levels x_1 and x_2 such that $x_2 > x_1$. The lower ability individual is in equilibrium at point A_1 where the budget line is tangent to his indifference curve. Since at every point, and at point A_1 in particular, the indifference curve of the higher ability individual is shallower, and since the budget line is positively sloped and concave, the tangency point of this individual occurs to the right of A_1, at a point such as A_2. Hence, for $\alpha \geqslant 0$, the rankings of abilities and of gross incomes are identical.

---

7. The difference in abilities shows up as a difference in tastes because of the focus on the individual's trade-off between consumption and gross income rather than on the trade-off between consumption and leisure for which all individuals possess the same indifference curves.

For $\alpha < 0$, normality of consumption is sufficient to establish that y and x are positively related. Performing a comparative statics experiment with respect to r on the first-order condition in (7), the resulting change in consumption is

$$\frac{dc}{dr} = \frac{1}{|D|}\{x[1 - T'(y)]u_{cl} - u_{ll} + 2\alpha x^2 u_c\}$$

Here D is the second-order condition for the problem in equation (6). Its explicit form appears in footnote 6. Performing a comparative statics experiment with respect to x and using this expression it can be shown that

$$\frac{dy}{dx} = \frac{1}{|D|}\left\{x[1 - T'(y)]u_c + n\left(|D|\frac{dc}{dr} - 2\alpha x^2 u_c\right)\right\}$$

Since $u_c > 0$, $\alpha < 0$, and consumption is a normal good ($dc/dr > 0$), gross income and ability are positively related for the case $\alpha < 0$ also.

For $\alpha \geq 0$ the convexity of the individual's budget set assures that given the parameters r, τ, α, and x there is only one value of n that solves the maximization problem in equation (6). However, for $\alpha < 0$, the budget set is not convex so that multiple solutions for n may occur. We rule such situations out by requiring that the second partial derivative of (6) with respect to n (whose explicit form is in footnote 6) is negative for all $0 \leq x \leq x_u$ and $0 \leq n \leq 1$. This requirement constitutes a joint restriction on the degree of concavity of the utility function and on the degree of regressivity of the tax schedule when this schedule is regressive.

III. The Political Process and the Government Budget Constraint

Choice of the parameters of the tax function is a political economy decision. In our analysis, the parameters of the tax schedule are determined by majority voting. Therefore only tax schedules that cannot be defeated through majority rule by *any* other schedule can be a political equilibrium. Voters are informed about their tastes and about the effects of taxes and redistribution on earned income implied by the model in the previous section. Taxes are levied to finance a fixed level of per capita government expenditure, G, and the amount of lump sum per capita redistribution in the form of demogrants that results from the political process.

Let $F(x)$ denote the distribution function of individual productivity so that $F(x)$ is the fraction of the population with productivity not greater than x. Budgetary deficits and surpluses are not possible, so the budget is balanced by taxes on the incomes of those who work:

$$\int_{x_0}^{x_u} \{\tau x n(x) + \alpha[x n(x)]^2\} dF(x) = G + r \qquad (13)$$

Since G is taken to be exogenous it is set to zero for simplicity.[8] Equation (13) implicitly determines r as a function of τ and α. This function is denoted

$$r = r(\tau, \alpha) \qquad (14)$$

Provided the utility function and the distribution function of individual ability are continuous, so is the function $r(\cdot)$. Following Romer (1975), we refer to (14) as the tax possibility frontier (TPF). Ignoring effects on incentives and income, an increase in either the flat component of the tax structure, τ, or in the degree of progressivity, α, increases revenues and the amount of per capita redistribution, r. When incentive effects are taken into account, an increase in either τ or α may decrease the tax base so much that redistribution must decrease to balance the budget. This is more likely if the tax burden is relatively high. The derivatives of r with respect to τ and α (denoted r_τ and r_α) may therefore be either positive or negative. The range along the TPF where either or both of r_τ and r_α are nonpositive is inefficient, since from any point in this range it is possible to increase redistribution without increasing tax burdens. Since *all* individuals like such changes, majority rule will never induce a political equilibrium along the inefficient range of the TPF. This is summarized in the following proposition.

Proposition 1: Under majority rule the voting equilibrium is never in the range of the TPF along which $r_\tau \leq 0$ or $r_\alpha \leq 0$, or both.

From equation (13), the derivatives of the TPF with respect to τ and α are[9]

$$r_\tau = \frac{1}{H} \int_{x_0}^{x_u} \left\{ y(x) + xT'[y(x)] \frac{\partial n}{\partial \tau}(x) \right\} dF(x) \qquad (15a)$$

$$r_\alpha = \frac{1}{H} \int_{x_0}^{x_u} \left\{ [y(x)]^2 + xT'[y(x)] \frac{\partial n}{\partial \alpha}(x) \right\} dF(x) \qquad (15b)$$

$$H \equiv 1 - \int_{x_0}^{x_u} xT'[y(x)] \frac{\partial n}{\partial r}(x) dF(x) \qquad (15c)$$

where $\partial n/\partial z(x)$, $z = \tau, \alpha, r$ are the responses of the labor supply of an individual with ability x to ceteris paribus changes in the tax parameters τ, α, and r. Assuming that a ceteris paribus increase in redistribution decreases individual labor supply (or does not increase it) $\partial n/\partial r \leq 0$. Since $T'[\cdot] \geq 0$, this implies that H is bounded away from zero and positive. Inspection of equations (15a) and (15b) reveals that, since $y(x)$, x, $\partial n/\partial \tau$, $\partial n/\partial \alpha$, and $T'[\cdot]$ are all bounded from above so are the numerators of these equations. Since H is bounded away from zero, this implies that r_τ and r_α are finite. Hence there is only one value of r that corresponds to each (τ, α) pair; the function $r(\tau, \alpha)$ is single valued.

---

8. The case in which there is an endogenously determined public good is briefly discussed in Section VIII.

9. The derivations are in part 1 of the appendix.

It is well known that majority rule induces collective intransitivities that can lead to cycles in the composition of the majority, when voters make multidimensional choices. This is an aspect of Arrow's (1951) impossibility theorem. In the contest of the present model, voters' preferences are affected by the three parameters r, τ, and α that define a tax schedule. The TPF in (14) reduces the dimensionality of the problem to the two parameters τ and α. But this alone does not ensure that there exists a unique tax schedule along the TPF that is preferred by a majority. In general there may be no Condorcet winner or, in the terminology of game theory, the core of the majority-rule game over tax schedules may be empty.

Proposition 1 implies that if there exists a majority-winning tax schedule, it must be in the range of the TPF in which $r_\tau > 0$ and $r_\alpha > 0$. The following discussion establishes conditions for the existence of a majority political equilibrium and characterizes it. Let $s \equiv (r, \tau, \alpha)$ be a tax schedule. Let S_m be the set of tax schedules along the TPF most preferred by the individual with median ability. This individual will be referred to as "the median individual" or simply "the median." Let s_m be a tax schedule in S_m. The set S_m may, but does not have to, include several tax schedules. Obviously when it does the median is indifferent between the different schedules. The following two definitions help to organize the discussion that follows.

Definition 1: A tax schedule s is a *local majority winner* if it is on the TPF and if there is no other schedule in the neighborhood of this schedule on the TPF that is strictly preferred by a majority.

Definition 2: A tax schedule is a *global majority winner* if it is on the TPF and if there is no other schedule on the TPF that is strictly preferred by a majority.

The local majority winner concept helps to characterize the conditions under which a global majority winner exists. In addition it clarifies the voters' choice when the political system contemplates only small changes in the status quo, as in Kramer and Klevorick (1974) and Romer (1977).

The following discussion, that culminates in theorem 1, demonstrates that all the schedules in the set S_m are local majority winners. Let

$$s_m \equiv (r_m, \tau_m, \alpha_m) \tag{16}$$

be any schedule in the set S_m. Let

$$dr \equiv r - r_m \qquad d\tau \equiv \tau - \tau_m \qquad d\alpha \equiv \alpha - \alpha_m \tag{17}$$

be a local deviation of s from s_m along the TPF. Let

$$I(r, \tau, \alpha; x) \equiv \max_n u[r + (1 - \tau)xn - \alpha(xn)^2, 1 - n] \tag{18}$$

be the indirect utility function of an individual with ability x. Since $u(\cdot)$ and the individual's budget constraint are continuous and the range of $u(\cdot)$ is compact, $I(\cdot)$ is a continuous function of the parameters r, τ, and α. In addition the implicit function theorem guarantees the differentiability of $I(\cdot)$ with respect to the three parameters.[10] Differentiating (18) totally with respect to a combined local change in the parameters of the tax schedule and using the envelope theorem

$$dI(y) = u_c(dr - y d\tau - y^2 d\alpha) \qquad (19)$$

Lemma 1: The schedule s_m is a local majority winner whenever either of the following holds

(i) $d\tau > 0$ and $d\alpha > 0$
(ii) $d\tau > 0$ and $d\alpha = 0$
(iii) $d\tau = 0$ and $d\alpha > 0$
(iv) $d\tau < 0$ and $d\alpha < 0$

Proof: Since $s_m \in S_m$, the median either dislikes or is indifferent to the change. Hence, from (19)

$$dI(y_m) = u_c^m(dr - y_m d\tau - y_m^2 d\alpha) \leq 0 \qquad (20)$$

where the subscript or superscript designates that the appropriate quantity refers to the median. In case (i), since $d\tau > 0$ and $d\alpha > 0$

$$dI(y) < dI(y_m) \leq 0 \quad \text{for all } y > y_m \qquad (21)$$

Since gross income is increasing in productivity, all the individuals with ability above that of the median dislike the change. If the median dislikes the change too there is a majority against it. If the median is indifferent there is a tie between those who like and those who dislike the change. In either case there is no schedule that is preferred to s_m by a majority. In cases (ii) and (iii) the inequality in (21) is still satisfied for all $y > y_m$. Hence the same considerations apply and there is no majority for the change.

In case (iv)

$$dI(y) < dI(y_m) \leq 0 \quad \text{for all } y < y_m \qquad (22)$$

Hence if the median dislikes the change there is a majority against it, and if the median is indifferent the majority does not prefer the change. In either case s_m is a local majority winner for the set of changes specified in the lemma. □

Lemma 2: The schedule s_m is a local majority winner against any change in schedule such that $d\tau < 0$, $d\alpha > 0$.

---

[10] Varian (1978), p. 267.

Proof: The change in welfare experienced by an individual with income y as a result of the shift from s_m to s is given by equation (19). Since $u_c > 0$ for all y, the change in welfare is positive, negative, or zero depending on whether the following second-degree polynomial in y is positive, negative, or zero:

$$P(y) \equiv -(d\alpha)y^2 - (d\tau)y + dr \tag{23}$$

The roots of $P(y)$ are

$$y_{c1} = \frac{1}{2}\left[-\frac{d\tau}{d\alpha} - \sqrt{\left(\frac{d\tau}{d\alpha}\right)^2 + 4\frac{dr}{d\alpha}}\right] \tag{24a}$$

$$y_{c2} = \frac{1}{2}\left[-\frac{d\tau}{d\alpha} + \sqrt{\left(\frac{d\tau}{d\alpha}\right)^2 + 4\frac{dr}{d\alpha}}\right] \tag{24b}$$

Since $-d\alpha < 0$, the polynomial $P(y)$ has a maximum and looks like an inverted U. If both roots are imaginary, $P(y)$ is either positive or negative for all $y - s$. When $P(y_m) < 0$, this implies that $P(y) < 0$ for all y, and all voters dislike the change. If both roots are real and distinct (see panel a of Figure 5-2)

$$P(y) < 0 \quad \text{for all } y > y_{c2} \text{ and } y < y_{c1} \tag{25a}$$

$$P(y) > 0 \quad \text{for all } y_{c1} < y < y_{c2} \tag{25b}$$

$$P(y) = 0 \quad \text{for } y = y_{c1} \text{ and } y = y_{c2} \tag{25c}$$

By assumption $P(y_m) < 0$, so y_m must be in the range defined by equation (25a). If $y_m > y_{c2}$, at least all individuals with incomes above y_m dislike the change too, and if $y_m < y_{c1}$, at least all individuals with incomes below y_m dislike the change. In either case there is a majority against the change.

When $P(y_m) = 0$, y_m is equal to either y_{c1} or y_{c2}. If y_{c1} and y_{c2} are distinct, the previous argument implies that there is always a majority against the change. If y_{c1} and y_{c2} collapse to one root, y_m equals its common value and $P(y)$ touches the horizontal axis only at this point. All other points are therefore below the horizontal axis, since $P(y)$ has a maximum at y_m. Hence everybody except for the median strictly dislikes the change. It follows that s_m is a local majority winner for all changes of the type $d\tau < 0$ and $d\alpha > 0$. □

Lemma 3: The schedule s_m is a local majority winner against any change in schedule such that

$$dr \leq 0 \qquad d\tau > 0 \qquad d\alpha < 0$$

Proof: When $d\alpha < 0$, $P(y)$ has a minimum and looks like a U (panel b of Figure 5-2). The value of $P(y)$ at the minimum is

$$\frac{1}{4}\frac{(d\tau)^2}{d\alpha} + dr \tag{26}$$

a. The case $d\tau < 0, d\alpha > 0$ (lemma 2)

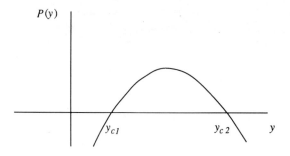

b. The case $dr \leq 0, d\tau > 0, d\alpha < 0$ (lemma 3)

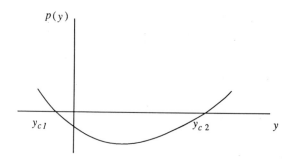

c. The case $dr > 0, d\tau > 0, d\alpha < 0$ (lemma 4)

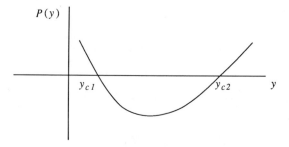

Figure 5-2.

which is negative, since $d\alpha < 0$ and $dr \leq 0$. Since for extreme values of y, $P(y)$ eventually becomes positive, the two roots—y_{c1} and y_{c2}—are real and distinct.

Since $dr \leq 0$ and $d\alpha < 0$ (24a) implies that $y_{c1} \leq 0$. Since $P(y_m) \leq 0$, $y_{c1} \leq y_m \leq y_{c2}$. Hence at least all the individuals with incomes below y_m weakly dislike the change. Since the median also weakly dislikes the change there is no majority in favor of the change and s_m is a local majority winner. □

From the TPF in (14)

$$dr = r_\tau d\tau + r_\alpha d\alpha \tag{27}$$

Dividing (27) by $d\alpha$ and substituting it into equations (24a) and (24b)

$$y_{c1}(a) = \frac{1}{2}(a - \sqrt{a^2 - 4r_\tau a + 4r_\alpha}) \tag{28a}$$

$$y_{c2}(a) = \frac{1}{2}(a + \sqrt{a^2 - 4r_\tau a + 4r_\alpha}) \tag{28b}$$

$$a \equiv -\frac{d\tau}{d\alpha} \tag{28c}$$

The only type of change in schedules not covered by lemmas 1 through 3 is the case $dr > 0$, $d\tau > 0$, $d\alpha < 0$. The next lemma demonstrates that if there is an internal solution along the TPF for s_m, this case can be disregarded.

Lemma 4: If the schedule s_m preferred by the median individual does not occur along the boundary[11] of the TPF, changes of the type $dr > 0$, $d\tau > 0$, $d\alpha < 0$ cannot occur.

Proof: Since $d\alpha < 0$ and $dr > 0$, $dr/d\alpha < 0$. It follows from equation (24a) that $y_{c1}(a) > 0$ for all a's in the range defined by the conditions of the lemma. Since $d\alpha < 0$, $P(y)$ has a minimum and can therefore be drawn as in panel c of Figure 5-2. Equation (27) implies that $dr > 0$ is equivalent to $a > r_\alpha/r_\tau$. Since $P(y_m) \leq 0$, $y_m < y_{c2}(a)$. Rearranging, this is equivalent to

$$a \leq \frac{r_\alpha - y_m^2}{r_\tau - y_m}$$

Hence we need to examine only cases for which a is in the range

$$\frac{r_\alpha}{r_\tau} < a \leq \frac{r_\alpha - y_m^2}{r_\tau - y_m} \tag{29}$$

---

11. The boundary of the TPF is defined by the restrictions $r > 0$ and $-\tau/2x_u \leq \alpha \leq (1 - \tau)/2x_u$. Hence the boundary is attained when $r = 0$ or $\alpha = -\tau/2x_u$ or $\alpha = (1 - \tau)/2x_u$.

In all other cases either dr is not positive or the median likes the change, contradicting the fact that s_m is weakly preferred by the median to all other schedules.

Since s_m occurs at an internal point along the TPF it must satisfy (applying the envelope theorem to (18)) the following two first-order conditions

$$\frac{dI_m}{d\alpha} = -u_c^m(\cdot)y_m\left(\frac{d\tau}{d\alpha} + y_m\right) = 0 \tag{30a}$$

$$\frac{dI_m}{dr} = u_c^m(\cdot)\left(1 - y_m\frac{d\tau}{dr}\right) = 0 \tag{30b}$$

where $d\tau/d\alpha$ is the change in τ resulting from a change in α along the TPF for a given r, and $d\tau/dr$ is the change in τ resulting from a change in r along the TPF holding α constant. For changing τ and α and constant r, equation (27) implies

$$\frac{d\tau}{d\alpha} = -\frac{r_\alpha}{r_\tau} \tag{31}$$

If the median works, $u_c^m(\cdot)y_m > 0$ and equations (30a) and (31) imply

$$y_m = \frac{r_\alpha}{r_\tau} \tag{32}$$

Substituting (32) into the upper limit of the interval in (29)

$$\frac{r_\alpha - y_m^2}{r_\tau - y_m} = \frac{r_\alpha - (r_\alpha/r_\tau)^2}{r_\tau - (r_\alpha/r_\tau)} = \frac{r_\alpha r_\tau^2 - r_\alpha^2}{r_\tau^3 - r_\alpha r_\tau} = \frac{r_\alpha}{r_\tau}$$

Hence the interval in (29) reduces to

$$\frac{r_\alpha}{r_\tau} < a \leqslant \frac{r_\alpha}{r_\tau} \tag{33}$$

implying that there is no value of a that simultaneously satisfies $dr > 0$ and $P(y_m) \leqslant 0$. Hence changes of the type $dr > 0$, $d\tau > 0$, $d\alpha < 0$ can be disregarded.

If the median does not work, taxes are set to yield the maximum redistribution for the TPF; s_m is characterized by the two first-order conditions $r_\alpha = r_\tau = 0$. It follows from equation (27) that $dr = 0$ for all $d\alpha$ and $d\tau$ when the median does not work. Hence changes of the type $dr > 0$ are not possible. □

Theorem 1: Let S_m be the set of tax schedules along the TPF such that there is no other tax schedule along the TPF that is strictly preferred by the individual with median ability. Let s_m be an element of S_m. If s_m does not occur on the boundary of the TPF, then s_m is a local majority winner.[12]

Proof: The proof is a direct consequence of lemmas 1 through 4.

Suppose that a given schedule, s_1, along the TPF is the status quo. Now an alternative schedule, s_2, is proposed. If s_2 is strictly preferred by a majority, it becomes the status quo schedule. In all other cases, including ties, the original status quo s_1 remains in place unless another schedule that is strictly preferred by a majority is found. The process ends when no schedule that is strictly preferred by a majority to the status quo can be found. Using this mechanism to vote on tax schedules, theorem 1 implies that, once a schedule $s_m \in S_m$ becomes the status quo, the process stops and s_m is adopted provided only that schedules in the neighborhood of the status quo are considered as alternatives at each stage. If S_m is not a singleton, this does not pin down which of the schedules in S_m will be adopted. But it limits the search for equilibrium schedules to the set that is most preferred by the individual with median ability. This characteristic of the local majority-winning set of schedules proves useful for the characterization of the equilibrium degree of progressivity in Section V.

IV. Conditions for the existence of a Globally Winning Tax Schedule

The result of theorem 1 can be extended to any (global) change in schedule along the TPF provided some additional conditions are satisfied. The following discussion leads to a statement of the precise conditions. Let

$$n(s, x), \quad y(s, x)$$

be the labor input and the gross labor income of an individual with productivity x when the tax schedule is s. Let

$$\Delta r \equiv r - r_m \qquad \Delta \tau \equiv \tau - \tau_m \qquad \Delta \alpha \equiv \alpha - \alpha_m \qquad (34)$$

be any global movement away from s_m along the TPF. The following lemma establishes that if an individual dislikes this change, his net income or consumption, with his employment fixed at $n(s_m, x)$, must be lower after the change.

---

12. The set S_m does not always contain all the local majority winning tax schedules. For example there may exist a tax schedule, s_l, along the TPF and not in S_m that is strictly preferred by the median individual to any schedule in the vicinity of s_l. This occurs when none of the schedules in S_m is in the vicinity of s_l. Since s_l does not belong to S_m, the median individual strictly prefers any of the schedules in S_m to s_l. Nonetheless s_l is a local majority winner, since the arguments leading to theorem 1 imply that if the median individual weakly prefers s_l to any schedule in the neighborhood of s_l, there is a weak majority in favor of s_l whenever s_l is confronted with any of the schedules in the vicinity of s_l. Hence the set of local majority winners may be larger than S_m. But (subject to the conditions of Section IV below) the set of global majority winners is identical to S_m since any of the (more distant) schedules in S_m is weakly preferred by a majority to s_l.

Lemma 5: If an individual with productivity x dislikes (or is indifferent to) the shift from $s_m = (r_m, \tau_m, \alpha_m)$ to $s = (r, \tau, \alpha)$, then, holding his labor input constant at $n(s_m, x)$, his net income or consumption is lower at s than at s_m.

Proof: Given that the individual's employment remains at $n(s_m, x)$, the change in his consumption is, from equation (5)

$$-\Delta\alpha[y(s_m, x)]^2 - \Delta\tau y(s_m, x) + \Delta r \tag{35}$$

Suppose the expression in (35) is nonnegative, contrary to the assertion of the lemma. Since leisure remains the same, the welfare of the individual obviously increases if (35) is positive. When the individual is allowed to adjust his labor input, welfare increases even further, since he is allowed to take advantage of additional substitution possibilities between labor and leisure that were not available when labor was frozen as $n(s_m, x)$. If the expression in (35) is zero, leisure and consumption are unchanged, so his welfare does not change. But once he is allowed to adjust his labor, welfare increases. Hence if the expression in (35) is nonnegative the individual's welfare increases as a result of the change, contradicting the assumption that the individual dislikes the change. It follows that given that labor is frozen at $n(s_m, x)$, consumption is lower at s than at s_m if the individual dislikes the change. □

An important consequence of lemma 5 is that if the median dislikes the switch from s_m to s, his net income at the prechange level of work must be reduced by the change.

Theorem 2 below presents sufficient conditions for the existence of a global majority winner (s_m) and characterizes the conditions. The statement of the theorem requires some preliminary discussion, to which we turn next. Let $P^g(y)$ be the polynomial in (23) with the local changes $(dr, d\tau, d\alpha)$ replaced everywhere by global changes—$(\Delta r, \Delta \tau, \Delta \alpha)$. Let

$$\bar{r}_\tau \equiv \frac{r(\tau, \alpha_m) - r(\tau_m, \alpha_m)}{\tau - \tau_m} \tag{36a}$$

$$\bar{r}_\alpha \equiv \frac{r(\tau, \alpha) - r(\tau, \alpha_m)}{\alpha - \alpha_m} \tag{36b}$$

Then by definition

$$\Delta r = \bar{r}_\tau \Delta \tau + \bar{r}_\alpha \Delta \alpha \tag{37}$$

Let y_{c1}^g and y_{c2}^g be the roots of the polynomial $P^g(y)$. The explicit form of those roots is the same as in equations (24), after adjusting for the change from local to the global changes denoted by "Δ." In the case $\Delta\alpha < 0$, the polynomial has a minimum and is described qualitatively by either panel b or c of Figure 5-2.

Dividing (37) by $\Delta\alpha$ and substituting the result into the expressions for y_{c1}^g and y_{c2}^g we obtain

$$y_{c1}^g(b) = \frac{1}{2}(b - \sqrt{b^2 - 4\bar{r}_\tau b + 4\bar{r}\alpha}) \tag{38a}$$

$$y_{c2}^g(b) = \frac{1}{2}(b + \sqrt{b^2 - 4\bar{r}_\tau b + 4\bar{r}\alpha}) \tag{38b}$$

$$b \equiv -\frac{\Delta\tau}{\Delta\alpha} \tag{38c}$$

At prechange labor inputs, all individuals with gross labor incomes in the range

$$\max[0, y_{c1}] < y < y_{c2} \tag{39}$$

experience a drop in net income as a result of the change. But those who are just a little above $\max[0, y_{c1}]$ or just a little below y_{c2} may actually experience a small increase in welfare (in spite of the income drop) once they adjust their labor inputs optimally. Hence the set of prechange gross incomes at which welfare decreases as a result of the change is strictly contained in the set defined by (39). Condition (i) of theorem 2 below and the fact that $P^g(y)$ decreases as one moves away from y_{c1}^g and y_{c2}^g towards $b/2$ imply that the smaller set is convex. Hence there exist small but positive numbers $\varepsilon_1(a)$ and $\varepsilon_2(a)$ such that welfare decreases for all individuals with prechange gross income in the range

$$\max[0, y_{c1}^g(b) + \varepsilon_1(b)] < y < y_{c2}^g(b) - \varepsilon_2(b) \tag{40}$$

Given a tax schedule, an individual's level of labor input and therefore his gross income is determined by his productivity, x. Hence gross income of an individual with productivity x when the tax schedule is s_m can be written

$$y = y(s_m, x) \tag{41}$$

Let

$$x = \hat{x}(s_m, y) \equiv x\{y\} \tag{42}$$

be the inverse function to (41). Given s_m, it expresses x as a function of y. Since, given the tax schedule, the ranking of individuals by productivity coincides with their ranking by gross income, $x(y)$ is an increasing function of y.

Theorem 2: Let $\Delta r \equiv r - r_m$, $\Delta\tau \equiv \tau - \tau_m$, $\Delta\alpha \equiv \alpha - \alpha_m$ be any global change in the tax schedule along the TPF starting from a status quo at $s_m = (r_m, \tau_m, \alpha_m)$. Let the conditions of theorem 1 be satisfied. Then s_m is a global majority winner provided the following additional conditions are satisfied.

(i) If individual i dislikes the change, all individuals whose net income decreases by more than that of individual i, when employment is held fixed at prechange levels, dislike the change also.

(ii)
$$F\{x[y_{c2}^g(b) - \varepsilon_2(b)]\} - F(x\{\max[0, y_{c1}^g(b) + \varepsilon_1(b)]\}) \geq \frac{1}{2} \qquad (43)$$

for all b's that correspond to changes of the type $\Delta r \leq 0$, $\Delta \tau > 0$, $\Delta \alpha < 0$.

(iii)
$$F\{x[y_{c2}^g(b) - \varepsilon_2(b)]\} - F\{x[y_{c1}^g(b) + \varepsilon_1(b)]\} \geq \frac{1}{2} \qquad (44)$$

for all b's that satisfy

$$\frac{\bar{r}_\alpha}{\bar{r}_\tau} < b < \frac{r_\tau^2 - (r_\alpha/\bar{r}_\alpha)r_\alpha \bar{r}_\alpha}{r_\tau^2 - (r_\tau/\bar{r}_\tau)r_\alpha \bar{r}_\tau} \qquad (45)$$

and that correspond to changes of the type $\Delta r > 0$, $\Delta \tau > 0$, $\Delta \alpha < 0$.

Proof: In part 2 of the appendix.

Theorem 2 implies that every $s_m \in S_m$ is a global majority winner. In addition, since any tax schedule on the TPF but not in S_m is not strictly preferred by a majority, the set S_m contains *all* the majority-winning schedules.[13]

Some discussion of the conditions in theorem 2 is in order. Condition (i) is basically an implicit restriction on the possible set of utility functions. Condition (ii) is always satisfied when $y_{c1}^g(b) + \varepsilon_1(b) \leq 0$, since $0 \leq y_m \leq y_{c2}^g(b) - \varepsilon_2(b)$. The reason is that in all such cases at least half of the voters do not strictly prefer the change. This is more likely to be the case when $\varepsilon_1(b)$ is small. In turn, this will be the case when the utility difference between a compensating change in income that maintains the same level of net income and a compensating change that maintains the same level of utility is small. Condition (iii) needs to be satisfied only in a rather limited range. This condition requires that the density of individuals in the range of abilities around the median is sufficiently large. In more intuitive but less exact terms, it requires that the middle class be a sufficiently large fraction of society. Note that when $r_\alpha/\bar{r}_\alpha = r_\tau/\bar{r}_\tau$, the range of b over which this condition is required to hold is empty, so that the condition is not binding.[14] When $r_\alpha/\bar{r}_\alpha \neq r_\tau/\bar{r}_\tau$, the range in (45) is narrower (and the condition in (44) therefore less restrictive) the nearer are the ratios $r_\alpha/\bar{r}_\alpha$ and $r_\tau/\bar{r}_\tau$ to each other. Those ratios tend to be similar when the differences between the

---

13. In game theoretic terms this means that S_m and the core of the majority-rule game over tax schedules coincide.

14. In particular this condition is satisfied when $r_\alpha = \bar{r}_\alpha$ and $r_\tau = \bar{r}_\tau$ so that the TPF is a plane in the range between (τ_m, α_m) and (τ, α).

global and the local curvatures of the TPF in the τ and the α directions are of similar magnitudes.

V. Majority Rule and Income Tax Progressivity when the Decisive Voter Works

The previous two sections have established the decisiveness of the individual with median ability for the choice of tax schedule. Hence the choice of schedule under majority rule depends on the position of the median (or decisive) voter in the income distribution, and therefore on his level of ability or productivity. This section establishes conditions under which a decisive voter who works chooses to tax incomes at marginally progressive rates. The following section analyzes the tax schedule chosen by a voter who subsists on transfer payments and does not work.

The decisive voter's problem is to choose the parameters of the TPF by maximizing his indirect utility (18)

$$I[r, \tau, \alpha; x_m] \tag{46}$$

subject to the TPF in (14). Here x_m is the productivity of the person at the median of the ability distribution. Substituting (14) into (46), differentiating totally with respect to α and r, and using the envelope theorem, we obtain

$$\frac{dI_m}{d\alpha} = u_c(r_\alpha - y_m^2) \tag{47a}$$

$$\frac{dI_m}{d\tau} = u_c(r_\tau - y_m) \tag{47b}$$

The subscript m denotes the voter with median ability and income, so y_m is the pretax income of the decisive voter and I_m is his (indirect) utility. We focus on the case in which the tax schedule that is preferred by the median is not on the boundary of the TPF. In this case the parameters of s_m are obtained by equating equations (47a) and (47b) to zero. The resulting equations together with the TPF constitute three equations that, in principle, can be used to determine the three parameters r_m, τ_m, and α_m.[15]

We are less interested in explicit solutions for the three parameters than in the conditions under which majority rule results in rising marginal tax rates. Earlier studies that used a political economy framework, by Romer (1975), Roberts (1977), and Meltzer and Richard (1981, 1983), ruled out marginal progressivity by imposing a linear tax function. The tax function used here can accommodate either marginal progressivity ($\alpha > 0$), marginal regressivity

---

15. Similarly, corner solutions can be obtained by incorporating the constraints $r \geq 0$ and $-\tau/2x_u \leq \alpha \leq (1-\tau)/2x_u$.

($\alpha < 0$), or a constant marginal tax rate ($\alpha = 0$). We can derive a sufficient condition for marginal progressivity by requiring that the expression in (47a) be positive for all $\alpha \leq 0$. Using (15b) in (47a) and noting that $u_c > 0$, this is equivalent to the condition

$$\frac{\int_{x_0}^{x_u} \{[y(x, r, \alpha)]^2 + xT'[y(x, r, \alpha)]\partial n/\partial \alpha(x, r, \alpha)\} dF(x)}{1 - \int_{x_0}^{x_u} xT'[y(x, r, \alpha)]\partial n/\partial r(x, r, \alpha) dF(x)} > [y(x_m, r, \alpha)]^2 \text{ for all } \alpha \leq 0 \text{ and } \tau \quad (48)$$

The dependence of the various terms in (48) on the productivity class x and on the parameters r and α is recognized explicitly in the notation. By contrast, the parameter τ that is held constant in (47a), and therefore in (48), is subsumed into the functional forms.

Equation (48) assures that *all* (of the possibly many) equilibria are in the progressive range ($\alpha > 0$). The reason is that increases in α increase utility for *all* $\alpha \leq 0$, so the decisive voter will never stop at a nonpositive value of α. Using basic formulas for the mean and the variance and rearranging, the condition in (48) can be restated as

$$C(r, \alpha) \equiv [1 - F(x_0)]\{V(r, \alpha) + [\bar{y}(r, \alpha)]^2\} - [y(x_m, r, \alpha)]^2$$
$$+ \int_{x_0}^{x_u} xT'[y(x, r, \alpha)] \left\{ \frac{\partial n}{\partial \alpha}(x, r, \alpha) + [y(x_m, r, \alpha)]^2 \right.$$
$$\left. \times \frac{\partial n}{\partial r}(x, r, \alpha) \right\} dF(x) > 0 \quad \text{for all } \alpha \leq 0 \text{ and } \tau \quad (49)$$

Here $\bar{y}(\cdot)$ and $V(\cdot)$ are the mean and the variance of income of the working population. At $\alpha = 0$ (a linear tax schedule), condition (49) reduces to

$$C(r, 0) = [1 - F(x_0)]\{V(r, 0) + [\bar{y}(r, 0)]^2\} - [y(x_m, r, 0)]^2$$
$$+ \tau \int_{x_0}^{x_u} x \left\{ \frac{\partial n}{\partial \alpha}(x, r, 0) + [y(x_m, r, 0)]^2 \frac{\partial n}{\partial r}(x, r, 0) \right\} dF(x) > 0 \quad \text{for all } \tau$$
$$(49a)$$

Other things equal, conditions (49) and (49a) are more likely to be satisfied the larger the variance of income $V(\cdot)$ and the larger the mean income $\bar{y}(\cdot)$ in relation to median income $y(x_m, r, \alpha)$. Given α and r, $V(\cdot)$ is normally larger the larger the variance of abilities. The mean-median income spread is normally larger the larger is the degree of positive skewness in the distribution of abilities. This leads to the following proposition.

Proposition 2: Majority rule is more likely to produce a progressive tax structure the more spread out is the distribution of abilities and the larger the degree of positive skewness of this distribution.

Proposition 2 implies that, other things equal, voters are more likely to choose a progressive tax structure the larger the spread in the distribution of abilities and the more the distribution is skewed toward high productivity. The intuition underlying these results is familiar. Individuals with higher ability have a larger share of income than of votes. They earn higher incomes because their productivity is higher and, often, because they work more hours. Even if they work fewer hours, income and productivity are positively related in our model. Since the number of people with high productivity is relatively small, the median voter can readily form a majority that agrees to lower the flat tax rate τ and raise the degree of progressivity (or lower regressivity). The median voter chooses to do this if he can reduce his own tax burden without reducing the transfers he, and all others, receive from the budget.

A larger spread in the distribution of income increases the likelihood of marginal progressivity. The reason is that a wide spread of the distribution increases the tax base at relatively high incomes. The larger tax base increases the likelihood that a compensated increase in α will generate sufficient additional tax revenues to reduce the average tax paid by the decisive voter.

Effects on incentives modify the results just discussed. Income taxes alter labor-leisure choices and change the tax base. Some notion about the effects of incentives can be obtained by examining the conditions for $\alpha > 0$ when the first-order condition in (47b) for an optimal choice of τ is satisfied. Since $u_c > 0$, this condition implies $y_m = r_\tau$. Hence a sufficient condition for $\alpha > 0$ is, from (47a),

$$r_\alpha - r_\tau^2 > 0 \quad \text{for all } \alpha \leq 0 \tag{50}$$

Using equations (15) in (50), noting that $H > 0$, and rearranging, this is equivalent to

$$\int_{x_0}^{x_u} \left(y^2 + xT' \frac{\partial n}{\partial \alpha} \right) dF \left(1 - \int_{x_0}^{x_u} xT' \frac{\partial n}{\partial r} dF \right) - \left[\int_{x_0}^{x_u} (y + xT' \frac{\partial n}{\partial \tau}) dF \right]^2 > 0$$

$$\text{for all } \alpha \leq 0 \tag{51}$$

where the dependence on x and s has been suppressed for notational simplicity. Since $\partial n/\partial r < 0$

$$H \equiv 1 - \int_{x_0}^{x_u} xT' \frac{\partial n}{\partial r} dF > 0$$

Since r_α and r_τ are positive this implies that

$$\int_{x_0}^{x_u} \left(y^2 + xT' \frac{\partial n}{\partial \alpha} \right) dF \quad \int_{x_0}^{x_u} \left(y + xT' \frac{\partial n}{\partial \tau} \right) dF$$

are both positive. Since $\partial n/\partial \alpha$ and $\partial n/\partial \tau$ are both negative, this implies that a positive α is more likely the lower $|\partial n/\partial \alpha|$ and the higher $|\partial n/\partial \tau|$ and $|\partial n/\partial r|$ in the range $\alpha \leq 0$. This is summarized in the following proposition.

Proposition 3: Majority rule is more likely to produce a progressive tax structure the lower the disincentive effects of an increase in α and the higher the disincentive effects of an increase in either τ or r on employment.

The intuition underlying proposition 3 is the following. An increase in α can be used either to increase redistribution, r, or to decrease the flat component of the tax structure, τ, or to do a bit of both. To understand the intuition it is convenient to consider pure cases in which the increase in α is used either to reduce τ or to increase r. In the first case the tax burden on low-ability individuals is alleviated and the tax burden on high-ability individuals is increased. This is a good strategy for the median when the tax base is not decreased by much. This is the case if employment is relatively insensitive to the increase in α ($|\partial n/\partial \alpha|$ small) and relatively sensitive to the increase in τ ($|\partial n/\partial \tau|$ large). This will be the case, in turn, if the labor supply of high-income individuals is less sensitive to a change in the marginal rate of taxation than the labor supply of low-income individuals.

Consider now an increase in α that is used solely to increase redistribution r. As can be seen from equations (15a) and (15b), large values of $|\partial n/\partial r|$ reduce both r_α and r_τ because of the larger negative incentive effects of an increase in redistribution on work. But, as can be seen from (50), the median finds it worthwhile to increase α whenever r_α is larger than r_τ^2. Since $H > 1$, larger $|\partial n/\partial r|$ reduce r_τ^2 by more than they reduce r_α. The incentive to increase α is therefore larger when $|\partial n/\partial r|$ is larger.

Condition (49) is sufficient for global progressivity whereas condition (49a) assures progressivity only locally. Since condition (49a) is simpler it is useful to know under what circumstances this condition alone is sufficient for global progressivity. If $C(r, \alpha)$ is a decreasing function of α, condition (49a) alone is sufficient. The reason is that, for negative α's, $C(\cdot)$ is a fortiori positive. We turn therefore to a discussion of the channels through which an increase in α affects $C(\cdot)$.

An increase in α corresponds to a movement toward higher values of α and r along the TPF, keeping τ constant. The increase in α produces (at the original levels of income) an increase in the tax burdens of all working individuals so that they reduce their levels of work. The increase in r causes a further reduction in the labor input of all individuals. If $\partial n/\partial \alpha$ and $\partial n/\partial r$ do not differ much across individuals, the work levels of different individuals decrease by roughly similar amounts, but the incomes of abler individuals decrease by more. As a consequence, a movement toward higher values of α and r along the TPF produces a decrease in both the mean and the variance of income. The increase in α increases the tax burden of individuals with higher incomes relative to taxpayers as a group. If, as a result, $|\partial n/\partial \alpha|$ is larger for individuals with larger incomes, the downward effect on V and $\bar{y}$ is even stronger. The increase in r also raises the threshold productivity level below which individuals choose to remain idle. This raises x_0 and lowers $1 - F(x_0)$.

On the other hand, the increase in α and r by decreasing $y(x_m, r, \alpha)$ and by raising the lower limit of the integral in (49) tends to increase $C(\cdot)$ as α and r increase along the TPF. If the elasticity of individual labor supply with respect to the combined change in α and r is smaller than 1 in absolute value, $T'[\cdot]$ increases,[16] countering some of the effects just described by increasing the weights on the negative $\partial n/\partial \alpha$ terms. In sum, for negative values of α, when the negative effects through V, $\bar{y}$, $1 - F(x_0)$, and possibly $T'[\cdot]$ dominate any positive effects,

$$\frac{\partial C(r, \alpha)}{\partial \alpha} < 0 \quad \text{for } \alpha < 0 \tag{52}$$

When the condition in (52) is satisfied, (49a) is sufficient to ensure that $\alpha > 0$.

VI. Majority Rule and Tax Progressivity when the Decisive Voter Does Not Work

In most democracies, the decisive voter works. He may receive transfers, but he also pays taxes. His decision to impose marginal progressivity depends on the balancing of the gains and losses he experiences and thus on parameters of the distribution of income and the labor supply response of those who pay the highest marginal rates. A nonworker faces a simpler problem. He pays no taxes, so his interest in marginal progressivity is greater. His own welfare depends on the transfers he receives, but these transfers do not increase with marginal progressivity if disincentive effects on income earners are strong. Consequently, a rational decisive voter who does not work never chooses a value of α that lowers redistribution.

The utility of a decisive voter who does not work increases monotically with r. To maximize utility, he chooses the point on the TPF at which r is maximized. Formally, we can state his problem as

$$\max_{\tau, \alpha} r(\tau, \alpha) \tag{53}$$

The solution to this problem yields the following two first-order conditions.

$$r_\alpha(\tau, \alpha) = 0 \tag{54a}$$

$$r_\tau(\tau, \alpha) = 0 \tag{54b}$$

from which, together with (14), it is possible to solve in principle for the decisive voter's preferred tax schedule s_m. Using (15b) and the fact that H is positive, a sufficient condition for progressivity is

$$\int_{x_0}^{x_u} \left(y^2 + xT' \frac{\partial n}{\partial \tau} \right) dF > 0 \quad \text{for all } \alpha \leqslant 0 \text{ and } \tau \tag{55}$$

---

[16] The total change in $T'[\cdot]$ at income y is $2y(1 + \eta_{n\alpha}^T)$ where $\eta_{n\alpha}^T$ is the (negative) elasticity of labor supply with respect to the combined increase in α and r along the TPF. Hence if $|\eta_{n\alpha}^T| < 1$, $T'[\cdot]$ increases as a result of the change.

which is equivalent to

$$K(r, \alpha) \equiv [1 - F(x_0)](V + \bar{y}^2) - \int_{x_0}^{x_u} xT'|\frac{\partial n}{\partial \alpha}|dF > 0 \quad \text{for all } \alpha \leq 0 \text{ and } \tau. \tag{56}$$

At $\alpha = 0$ this condition reduces to

$$[1 - F(x_0)](V + \bar{y}^2) - \tau \int_{x_0}^{x_u} x \left|\frac{\partial n}{\partial \alpha}\right| dF > 0 \quad \text{for all } \tau \tag{56a}$$

We saw in the previous section that when, given τ, we move in the direction of a higher α and a higher r along the TPF $(1 - F(x_0))(V + \bar{y}^2)$ goes down. Hence if the last term in (56) does not go down, $K(r(\alpha, \tau), \alpha)$ is (given τ) a decreasing function of α. This leads to the following proposition.

Proposition 4: If the median does not work and $A(r, \alpha) \equiv \int_{x_0}^{x_u} xT'|\partial n/\partial \alpha| dF$ is a nondecreasing function of α for all $0 \leq \tau \leq 1$, condition (56a) is sufficient to ensure marginal progressivity.

$A(r, \alpha)$ will be nondecreasing in α if $|\eta_{n\alpha}^T| < 1$ for all x and if the reduction in $A(\cdot)$ due to the increase in x_0 does not dominate the increase in $A(\cdot)$ because of the increase in T'. The previous statement implicitly assumes that $|\partial n/\partial \alpha|$ does not depend on α. If $|\partial n/\partial \alpha|$ increases with α (implying that the marginal disincentive effect of α on work grows as the tax burden increases), $A(\cdot)$ is more likely to increase with α.

Proposition 4 confirms the intuition that some of the factors that induce marginal progressivity are the same whether or not the median works. In particular, the larger the variance of abilities, the more skewed to the right is their distribution; and the smaller the disincentive effects of an increase in α, the more likely that majority rule will produce a marginally progressive tax schedule.

Finally, comparison of conditions (49) and (49a) with conditions (56) and (56a), respectively, suggests that if a progressive tax schedule arises when the median works it must arise a fortiori when the median does not work.

VII. A Remark on Progressivity in the Presence of Public Good

Since individual utility from expenditures on public goods has not been modeled explicitly, one may get the erroneous impression that progressivity of the tax schedule arises only when the budget is used for redistributional purposes. In fact the same elements that are conducive to progressivity when the budget is used to redistribute income are likely to lead to progressivity when it is used to provide a public good. A general demonstration of this claim is beyond the scope of this chapter. Instead we illustrate it for a particular utility function in the case in which the entire budget is used to finance a public good.

Let g be the amount of a public good enjoyed by a representative individual and let

$$v(c + g, l) \tag{57}$$

be the utility function of a typical individual. This specification, which implies that c and g are perfect substitutes, is adopted for simplicity. The TPF in (13) is replaced by

$$g = \delta N \int_{x_0}^{x_u} \{\tau x n(x) + \alpha [x n(x)]^2\} dF(x) \equiv \delta N r \tag{13a}$$

where N is the number of individuals in the economy and δ is a parameter between $1/N$ and 1. When $\delta = 1$, g is a pure public good. When $\delta = 1/N$, g is a publicly provided private good. In the more likely intermediate cases, $1/N < \delta < 1$, g is a public good but not a pure public good. Private consumption is now given by

$$c(x) = (1 - \tau)xn - \alpha(xn)^2 \tag{5a}$$

For $\delta = 1/N$, equation (13a) implies that $g = r$, so the model of this section becomes formally identical to the model of the previous sections. Hence the same factors that were conducive to progressivity before are conducive to progressivity now as well.

For $\delta > 1/N$, g is larger than r by some fixed factor. As a result, all individuals prefer a larger budget than in the case $\delta = 1/N$. The reason is that the marginal utility of the public good is now higher. However, the same conflicts of interest regarding the *financing* of g that existed when $\delta = 1/N$ are also present when $\delta > 1/N$. In particular the analysis of Sections III through VI can be replicated with r replaced by $\delta N r$. The formal conditions for the existence of a majority-winning schedule and for progressivity have to be adjusted to reflect this change. However, the qualitative results of propositions 2 and 3 are likely to carry over to this case too. The intuitive reasoning underlying this statement relies on the observation that an increase in the budget increases the consumption of the public good by the same amount for everybody, as was the case for $\delta = 1/N$. But the necessary increase in financing triggers redistributional conflicts that are essentially identical to those that are present when $\delta = 1/N$ since, except for δ, the model is the same.

This example suggests that, at least for some classes of utility functions, the factors that are conducive to progressivity when the budget is used to redistribute income are also conducive to progressivity when the budget is used solely to provide a public good.

VIII. Concluding Comments

Our intent in this chapter is to develop a positive theory of income taxation that generates tax schedules exhibiting marginal and average tax rates that rise with income. This feature is commonly found in many democratic countries, and in some states of the United States.

Economists and others have long speculated on the desirability of progressive taxes. Efforts to use theories of optimal taxation to explain the existence of progressivity have not been completely successful. Maximization of a Benthamite criterion very seldom leads to progressive tax structures. An alternative view, taken here, is that tax schedules are the outcome of a political equilibrium in which the majority imposes its will on the minority. All self-interested individuals would like to pay no taxes and to obtain positive redistribution from the government. Obviously this is not feasible for everybody. But in a democratic society in which tax schedules are determined by majority rule, the low- and middle-income majority can impose a certain level of redistribution on the more affluent minority. The analysis provides conditions under which this type of democracy leads to progressivity in the taxation of income.

A fundamental problem that arises once tax schedules are specified in a way that is sufficiently flexible to allow progressivity is that a majority-winning tax schedule need not exist. We derive conditions for the existance of a winning schedule within the set of quadratic schedules. An important condition for existence is that the ranking of gross incomes and of abilities be the same for all tax schedules. A sufficient condition for such identical ranking is the normality of both consumption and leisure. Under this condition, with some additional restrictions on the utility function and on the distribution of abilities, the individual with median ability is shown to be decisive. This result paves the way for finding conditions for progressivity, since it reduces this task to that of finding conditions under which the individual with median ability prefers progressivity. In contrast to the optimal taxation literature, we find that the set of circumstances under which the decisive voter, who maximizes utility, imposes progressivity is nonnegligible. Marginal progressivity is more likely (1) the larger the spread of the distribution of abilities in the population, (2) the smaller the labor supply response of the relatively more productive to an increase in marginal tax rates, and (3) the larger the difference between the ability of the decisive (median) voter and the mean ability of the community. (4) When the median works, larger disincentive effects of redistribution on labor force participation also increase progressivity. These conditions do not have to be satisfied separately; their combined effect is sufficient.

The methodology used to demonstrate the existence of a local majority-winning tax schedule can probably be used to extend this result to any tax schedule by viewing the quadratic schedule as a second-order Taylor expansion of a more general class of schedules. But we have not done that.

The tax schedules that result from our analysis reflect, mainly, skewness of the distribution of income that puts average income above median income, the variance of the distribution, and the effects of tax rates and redistribution on incentives to work. If these factors were identical across countries, and the franchise approximately the same, we would predict common tax schedules in democratic countries. Differences in tax schedules principally reflect possible differences in the voting rule that determines the franchise, differences in the distribution of income and ability, and differences in the effect of incentives, as measured in our analysis by the marginal effect of tax rates on labor supply.

Although the focus of the paper is on deriving conditions for progressivity when the budget is used to redistribute income, it is likely that similar conditions are conducive to progressivity when the budget is used to provide a public good. As illustrated in Section VII, even the provision of a pure public good is not independent of redistributional considerations, because of the need to decide on ways to finance it.

Some limitations of the analysis should be noted. Our model is static; income must be interpreted as lifetime income.[17] The budget is always balanced. We neglect migration and other open economy considerations that can limit progressivity over the time frame to which our model is most applicable. There is no capital, and therefore there are no taxes on capital.

Despite these limitations, the political economy model appears to be useful for understanding the determination of tax schedules and for showing that majority rule implies marginal progressivity to finance income redistribution and other expenditures under a relatively wide set of circumstances. The median or decisive voter chooses rising marginal tax rates if he can thereby reduce his own taxes without lowering the transfers he receives, or if he can increase the transfers he receives without increasing his own tax rates. Casual observations for many economies suggest that the voting process produces an outcome of this kind.

Appendix

1. *Derivation of Equations (15)*

Totally differentiating (13) with respect to τ, holding α constant, we obtain

$$\int_{x_0}^{x_u} \left[y + \tau x \left(\frac{\partial n}{\partial \tau} + r_\tau \frac{\partial n}{\partial r} \right) + 2\alpha x^2 n \left(\frac{\partial n}{\partial \tau} + r_\tau \frac{\partial n}{\partial r} \right) \right] dF$$

$$- r_\tau - \{\tau x_0 n(x_0) + \alpha [x_0 n(x_0)]^2\} \left(\frac{\partial x_0}{\partial \tau} + r_0 \frac{\partial x_0}{\partial r} \right) = 0$$

---

17. A recent application of the majority-rule paradigm to intertemporal redistribution and the determination of the public debt and deficits appears in Cukierman and Meltzer (1989), represented as Chapter 6 of this volume.

where the dependence of the various terms on x is not made explicit in the notation. Since $n(x_0) = 0$, the last term drops out. Equation (15a) follows by rearranging and by using equation (4) in the text.

Totally differentiating (13) with respect to α, holding τ constant, we obtain

$$\int_{x_0}^{x_u} \left[y^2 + \tau x \left(\frac{\partial n}{\partial \alpha} + r_\alpha \frac{\partial n}{\partial r} \right) + 2\alpha x^2 n \left(\frac{\partial n}{\partial \alpha} + r_\alpha \frac{\partial n}{\partial r} \right) \right] dF$$

$$- r_\alpha - \{\tau x_0 n(x_0) + \alpha [x_0 n(x_0)]^2\} \left(\frac{\partial x_0}{\partial \tau} + r_\tau \frac{\partial x_0}{\partial r} \right) = 0$$

Equation (15b) follows by noting that $n(x_0) = 0$, using (4), and by rearranging.

2. Proof of Theorem 2 on the Existence and Characterization of a Global Median.

It is convenient to break the proof into several lemmas.

Lemma A1: If condition (i) of theorem 2 is satisfied, s_m is a global majority winner against any change in tax schedule such that α and τ change in the same direction or such that only one of either α or τ changes.

Proof: Since $s_m \in S_m$, the median either dislikes or is indifferent to the change. Hence by lemma 5

$$P(y_m) \equiv -\Delta \alpha y_m^2 - \Delta \tau y_m + \Delta r < 0 \tag{A1}$$

where

$$y_m \equiv y(s_m, x_m)$$

Multiplying (A1) by $-1/\Delta \alpha$

$$Q(y_m) \equiv y_m^2 + \frac{\Delta \tau}{\Delta \alpha} y_m - \frac{\Delta r}{\Delta \alpha} \tag{A2}$$

(A1) and (A2) imply

$$Q(y_m) < 0 \quad \text{if } \Delta\alpha < 0 \tag{A3a}$$

$$Q(y_m) > 0 \quad \text{if } \Delta\alpha > 0. \tag{A3b}$$

Note that

$$Q'(y) = 2y + \frac{\Delta\tau}{\Delta\alpha} \tag{A4}$$

which is positive for all $y \geq 0$ when $\Delta\tau$ and $\Delta\alpha$ have the same signs. Hence

$$Q(y) < Q(y_m) \quad \text{for all } \Delta\alpha < 0 \text{ and all } y < y_m \quad \text{(A5a)}$$
$$Q(y) > Q(y_m) \quad \text{for all } \Delta\alpha > 0 \text{ and all } y > y_m \quad \text{(A5b)}$$

(A3) and (A5) imply

$$P(y) < P(y_m) \quad \text{for all } \Delta\alpha < 0 \text{ and all } y < y_m \quad \text{(A6a)}$$
$$P(y) < P(y_m) \quad \text{for all } \Delta\alpha > 0 \text{ and all } y > y_m \quad \text{(A6b)}$$

But $P(y)$ is the change in the net income of an individual with prechange income y when he is not allowed to adjust his labor income. Condition (i) of theorem 2 therefore implies that

$$\text{when } \Delta\alpha > 0, \ s_m P_i s \quad \text{by all } y_i < y_m \quad \text{(A7a)}$$
$$\text{when } \Delta\alpha < 0, \ s_m P_i s \quad \text{by all } y_i > y_m \quad \text{(A7b)}$$

The notation $s_m P_i s$ should be read, "s_m is prefferred to s by an individual with prechange income y_i." Since the ranking of gross incomes is the same as that of abilities, (A7) implies that there is no change in schedule, such that $\Delta\alpha$ and $\Delta\tau$ have the same sign, that is strictly preferred by a majority.

When $\Delta\tau = 0$ and $\Delta\alpha \neq 0$, the fact that $s_m \in S_m$ in conjunction with lemma 5 implies

$$P(y_m) = -\Delta\alpha y_m^2 + \Delta r < 0$$

This implies

$$P(y) < P(y_m) \quad \text{for all } \Delta\alpha < 0 \text{ and all } y < y_m \quad \text{(A8a)}$$
$$P(y) < P(y_m) \quad \text{for all } \Delta\alpha > 0 \text{ and all } y > y_m \quad \text{(A8b)}$$

When $\Delta\tau \neq 0$ and $\Delta\alpha = 0$, the fact that $s_m \in S_m$ in conjunction with lemma 5 implies

$$P(y_m) = -\Delta\tau y_m + \Delta r < 0$$

This implies

$$P(y) < P(y_m) \quad \text{for all } \Delta\tau < 0 \text{ and all } y < y_m \quad \text{(A9a)}$$
$$P(y) < P(y_m) \quad \text{for all } \Delta\tau > 0 \text{ and all } y > y_m \quad \text{(A9b)}$$

(A8), (A9), and condition (i) of theorem 2 imply that there is no majority that strictly prefers changes of the type $\Delta\alpha = 0$ and $\Delta\tau \neq 0$ or $\Delta\alpha \neq 0$ and $\Delta\tau = 0$.

□

Lemma A2: If condition (i) of theorem 2 is satisfied, s_m is a global majority winner against any change in schedule of the type

$$\Delta\alpha > 0, \quad \Delta\tau < 0$$

Proof: The fact that $s_m \in S_m$ and lemma 5 imply that (A1) holds. Since $\Delta\alpha > 0$, this is equivalent in turn to

$$Q(y_m) > 0 \qquad (A10)$$

Since $\Delta\tau/\Delta\alpha < 0$, equation (A4) implies

$$Q'(y) = \begin{cases} >0 & y > \dfrac{1}{2}\left|\dfrac{\Delta\tau}{\Delta\alpha}\right| \\ =0 & y = \dfrac{1}{2}\left|\dfrac{\Delta\tau}{\Delta\alpha}\right| \\ <0 & y < \dfrac{1}{2}\left|\dfrac{\Delta\tau}{\Delta\alpha}\right| \end{cases}. \qquad (A11)$$

(A10) and (A11) imply

$$Q(y) > Q(y_m) \quad \text{for all } y > y_m \text{ if } y_m > \dfrac{1}{2}\left|\dfrac{\Delta\tau}{\Delta\alpha}\right| \qquad (A12a)$$

$$Q(y) > Q(y_m) \quad \text{for all } y < y_m \text{ if } y_m < \dfrac{1}{2}\left|\dfrac{\Delta\tau}{\Delta\alpha}\right| \qquad (A12b)$$

$$Q(y) > Q(y_m) \quad \text{for all } y \neq y_m \text{ if } y_m = \dfrac{1}{2}\left|\dfrac{\Delta\tau}{\Delta\alpha}\right| \qquad (A12c)$$

Since $Q(y)$ and $P(y)$ are inversely related for $\Delta\alpha > 0$, this implies that in all three cases at least 50 percent of the voters suffer (at prechange labor inputs) a decrease in net income that is larger than $P(y_m)$. Condition (i) of theorem 2 implies therefore that at least the same number of voters dislike changes of the type $\Delta\alpha > 0$, $\Delta\tau < 0$. Since the median either dislikes or is indifferent to the change, there is no change of the type $\Delta\alpha > 0$, $\Delta\tau < 0$ that is preferred by a majority to s_m. Hence s_m is a global majority winner against changes of this type. □

Lemma A3: s_m is a global majority winner against any change in schedule of the type

$$\Delta r \leq 0, \qquad \Delta\tau > 0, \qquad \Delta\alpha < 0$$

if conditions (i) and (ii) of theorem 2 are satisfied.

Proof: Due to condition (i) of theorem 2, all individuals with gross incomes in the open segment defined by (40) weakly dislike the change. Condition (ii) of theorem 2 ensures that for any b such that $\Delta r \leq 0$, $\Delta \tau > 0$, $\Delta \alpha < 0$ there is a majority that weakly dislikes the change. □

Lemma A4: s_m is a global majority winner against any change in schedule of the type

$$\Delta r > 0, \quad \Delta \tau > 0, \quad \Delta \alpha < 0$$

if conditions (i) and (iii) of theorem 2 are satisfied.

Proof: Since $\Delta r > 0$ and since $P(y_m) < 0$, so that $y_m < y_{c2}^g$, an argument similar to that which led to equation (29) in the text implies that

$$\frac{\bar{r}_\alpha}{\bar{r}_\tau} < b < \frac{\bar{r}_\alpha - y_m^2}{\bar{r}_\tau - y_m} \tag{A13}$$

provided the median is employed. Since s_m does not occur on the boundary of the TPF, $y_m = r_\alpha/r_\tau$ (equation (32)). Hence the condition in (A13) reduces to the restriction on b in condition (iii) of theorem 2. Since $\Delta r > 0$, $y_{c1}^g > 0$. Condition (i) of theorem 2 implies that all individuals with prechange gross incomes in the range

$$y_{c1}^g(b) + \varepsilon_1(b) < y < y_{c2}^g(b) - \varepsilon_2(b) \tag{A14}$$

dislike the change. Condition (iii) of theorem 2 implies that for all b's in the open interval defined by (A14) there is no majority in favor of the change. Hence s_m is a global majority winner against all changes of the type $\Delta r > 0$, $\Delta \tau > 0$, $\Delta \alpha < 0$, when the median works.

If the median does not work, the fact that he does not prefer the change implies that redistribution, r, goes down or does not change as a consequence of the change in tax schedule. Hence the case $\Delta r > 0$, $\Delta \tau > 0$, $\Delta \alpha < 0$ is not possible when the median does not work, so there is no need to consider it. □

The proof of theorem 2 is completed by combining lemmas A1 through A4.

References

Arrow, K. J. (1951). *Social Choice and Individual Values*. New York: John Wiley.
Atkinson, A. B. (1973). "How Progressive Should Income Tax Be?" *Essays in Modern Economics*. London: Longmans. Reprinted in A. B. Atkinson (1983). *Social Justice and Public Policy*. Sussex, England: Wheatsheaf Books.
Blum, W. and Kalven, H. Jr. (1953). *The Uneasy Case for Progressive Taxation*. Chicago: University of Chicago Press.

Campbell, C. D. (ed.) (1977). *Income Redistribution*. Washington, D.C.: American Enterprise Institute.

Cukierman, A. and Meltzer, A. H. (1989). "A Political Theory of Government Debt and Deficits in a Neo Ricardian Framework." *American Economic Review*, 79 (Sept.), pp. 713–32.

Kramer, G. H. and Klevorick, A. (1974). "Existence of a 'Local' Cooperative Equilibrium in a Class of Voting Games." *Review of Economic Studies*, 41, pp. 539–47.

Kramer, G. H. and Snyder, J. M. (1983). Fairness, Self Interest, and the Politics of the Progressive Income Tax. SS-WP 498 California Institute of Technology, Pasadena (Nov.).

Kramer, G. H. and Snyder, J. M. (1984). Linearity of the Optimal Income Tax: A Generalization. SS-WP 534, California Institute of Technology, Pasadena (Mar.).

Meltzer, A. H. and Richard, S. F. (1981). "A Rational Theory of the Size of Government." *Journal of Political Economy*, 89 (Oct.) pp. 914–27.

Meltzer, A. H. and Richard, S. F. (1983). "Tests of a Rational Theory of the Size of Government." *Public Choice*, 41, pp. 403–18.

Mirrlees, J. A. (1971). "An Exploration in the Theory of Optimum Income Taxation." *Review of Economic Studies*, 38 (Apr.), pp. 175–208.

Phelps, E. S. (1973). "Taxation of Wage Income for Economic Justice." *Quarterly Journal of Economics*, 87 (Aug.), pp. 331–54.

Pigou, A. C. (1947). *A Study in Public Finance*, 3rd ed. New York: Macmillan.

Roberts, K. W. S. (1977). "Voting over Income Tax Schedules." *Journal of Public Economics*, 8, pp. 329–40.

Romer, T. (1975). "Individual Welfare, Majority Voting, and the Properties of a Linear Income Tax." *Journal of Public Economics*, 4, pp. 163–85.

Romer, T. (1977). "Majority Voting on Tax Parameters—Some Further Results." *Journal of* Public Economics, 7, pp. 127–33.

Sadka, E. (1976). "On Income Distribution, Incentive Effects, and Optimal Income Taxation." *Review of Economic Studies*, 43, pp. 261–67.

Sheshinski, E. (1972). "The Optimal Linear Income Tax." *Review of Economic Studies*, 39, pp. 297–302.

Simons, H. (1938). *Personal Income Taxation*. Chicago: Univ. of Chicago.

Varian, H. R. (1978). *Microeconomic Analysis*. New York: Norton and Company.

6

A Political Theory of Government Debt and Deficits in a Neo-Ricardian Framework

ALEX CUKIERMAN AND ALLAN H. MELTZER

Government expenditure is used for two main purposes—the provision of public goods and the redistribution of income. We focus here on the implications of redistribution for the size of the public debt, budgetary deficits, and surpluses. Since the focus is on redistribution, we abstract from the function of government as a provider of public goods and from issues that relate to minimization of the deadweight loss of taxation over time. This chapter can be viewed as complementary to the work of Robert Barro (1979), who proposed and tested a theory of public debt based on society's attempt to minimize the excess burden of taxation over time.

The main function of public debt is to redistribute the burden of taxation over time and across generations. In a neo-Ricardian world, such activity seems an idle exercise. Barro showed that, in the presence of an operative bequest motive and a perfect capital market, individuals totally undo the effects of debt-induced redistribution on consumption and welfare by adjusting their bequests appropriately. The existence of government debt in countries with developed capital markets, and the frequently stated belief that debt is a burden, is puzzling. Why do rational individuals complain about a burden that, according to Barro, does not occur?

The puzzle vanishes when individuals differ in abilities and therefore in wage earnings, and perhaps also in their initial nonhuman wealth. The reason is that some do not desire to leave positive bequests, and some would choose to borrow resources from future generations. As Allan Drazen (1978, footnote 1) has noted, individuals cannot obligate the future labor income of their descendants within

We would like to acknowledge very helpful discussions with Scott Richard, the comments of Alberto Alesina, and the suggestions of anonymous referees. A previous version of this chapter was presented at the Fourth Pinhas Sapir Conference on *Economic Effects of the Government Budget*, December 1986, Tel-Aviv University.

the existing institutional structure of democratic societies. The minimum bequest is constrained to zero; negative bequests are forbidden and, perhaps more important, do not have to be discharged. We refer to people who would choose to leave negative bequests as *bequest-constrained* individuals. These individuals favor any fiscal policy that increases their lifetime income at the expense of future generations even when the present value of the tax change is zero. For example, increased social security benefits financed by debt issues shift taxes forward and enable bequest-constrained individuals to achieve a superior allocation of consumption across generations within the same family.

Whether the bequest-constrained individuals succeed in pushing fiscal policy toward lower current taxes, higher debt, and higher social security benefits depends on the characteristics of the political process. Here we adopt the hypothesis that fiscal policy in democratic societies is determined by majority rule.[1] Hence, if decisive voters are bequest constrained, they will choose lower current taxes financed by additional debt and perhaps also increased social security benefits.

If decisive voters are not bequest constrained, they are indifferent to a reallocation of taxes and social security over time that has neutral present value. This indifference, which is a direct consequence of the Barro (1974) debt neutrality theorem, holds only if the reallocation does not affect wage rates and the return to capital. If some voters are bequest constrained, a present value preserving substitution of taxes for debt increases the consumption of bequest-constrained individuals. They obtain the additional resources from the non-bequest-constrained individuals who substitute bonds for capital in their portfolios. Although debt and capital are perfect substitutes in portfolios, they are not perfect substitutes in production. Additional debt crowds out some capital[2] and causes changes in returns to factors of production. Obviously, individuals are not indifferent to such induced general equilibrium effects of a higher debt, even if they are not bequest constrained.

There are two cases in which individuals with an operative bequest motive remain indifferent to the size of the government debt. First, if there are no bequest-constrained individuals in the economy, debt does not crowd out capital. Second, if a change in the capital/labor ratio does not affect factor returns in the relevant range, non-bequest-constrained individuals behave as in Barro (1974). In the latter case, however, voters relax the bequest constraint. The individual with the most severe bequest constraint is decisive; majority rule leads to a choice of debt that frees all bequest-constrained individuals from their constraints (provided the capital stock is sufficiently large).

In the general case, crowding out of capital affects the welfare of all individuals by changing factor returns. The rate of return to capital increases with the debt and, if labor and capital are complements in production, real

---

1. In this we follow Romer (1975), Roberts (1977), and Meltzer and Richard (1981), who all use this paradigm to characterize taxation and redistribution in a temporal framework.

2. The amount of capital that is crowded out by an additional unit of debt depends on the fraction of bequest-constrained individuals in the economy and on the extent to which they are constrained.

wages fall. Individuals for whom labor income is the major source of income are adversely affected by the increase in debt. If they are not bequest constrained, they vote against the increase. If they are bequest constrained, their vote depends on a comparison of the gain from intergenerational reallocation of consumption and the loss of welfare from the decrease in wages.

A main purpose of this chapter is to identify economic conditions that induce a larger debt and larger current deficits under majority rule. Debt is larger: (1) the larger is the size of the spread of the distribution of wealth (human and nonhuman) across individuals and the smaller the fraction of individuals for whom labor income is the main source of income; (2) the higher is the rate of technical progress; and (3) the less responsive are wages, and the more responsive the return to capital, to a change in the capital/labor ratio.

Majority choice of a large current debt does not necessarily imply a current deficit. Whether the current choice of debt level implies a current deficit or a surplus depends on whether the debt currently chosen is larger or smaller than the debt chosen by majority rule in the previous period. Deficits are likely when the decisive voter experiences a change in economic conditions that induces a larger debt. In particular, the likelihood of deficits increases with an increase in the rate of economic growth, an extension of the franchise to low wealth individuals who are likely to be bequest constrained, and an increase in the proportion of individuals whose main source of income is from returns to capital.

To bring out the reasons why debt and taxes are not equivalent in our model, we retain many of the features that are standard in this literature, including the overlapping generations model, due to Paul Samuelson (1958), lump-sum taxes, and Barro's (1974) intergenerational transfers from parent to child. There is no uncertainty. We differ from this standard framework by introducing, as in Allan Meltzer and Scott Richard (1981), differences in individual ability and therefore in wage rates. In addition, individuals receive different bequests. The position of each individual in the distribution of wealth, his wage rate, and the wage rates he expects for future generations in his family determine his attitude toward the size of the debt. Given individual preferences, majority rule determines the debt size and the current taxes chosen by voters. In addition, as in Peter Diamond (1965), the model explicitly recognizes the productive functions of capital and labor.

The chapter is organized as follows. The economy and the political institutions are described briefly in Section I. Section II shows the consumption, saving, and bequest decisions of bequest-constrained and non-bequest-constrained individuals for given tax and social security structures. It also summarizes conditions on wages, taxes, and bequests received under which an individual is likely to be bequest constrained. Section III describes more fully the institutional structure within which political decisions about debt, taxes, and social security benefits are made each period. It also characterizes the factors that determine individual attitudes toward intertemporal reallocations of taxes in the absence of general equilibrium effects. The core section of the paper, Section IV, combines results from previous sections to characterize the determination of public debt by majority rule. An important intermediate step derives individual attitudes toward debt in the presence of general equilibrium effects. Section V uses the results to derive conditions that are conducive to large debts and deficits. This section contains the main, empirically testable, implications of the theory. Concluding remarks follow.

I. Structure of the Economy and of the Political Process—Preliminaries

The Private Economy

The economy is represented by an overlapping-generations structure with bequests. Generation t is young in period t and old in period $t + 1$. The young of generation t overlap with the old of generation $t - 1$ in period t. Population is stationary. The number of young and old, denoted N, is identical across periods. Individuals work only when young. Each young individual differs in productivity. Each supplies inelastically one unit of labor per period, and each receives the wage for his type of labor. Output is produced by means of a constant returns-to-scale aggregate production function

$$F(G^t N, K_t) \qquad (1)$$

where K_t is the aggregate quantity of capital in period t, and G is the (gross) rate of labor augmenting technological progress. The capital and labor markets are competitive. The real rate of interest, r_t, is determined by the marginal productivity of capital and is given by

$$r_t = F_K(G^t N, K_t) \qquad (2)$$

The average wage rate in period t, $\bar{w}_t$, is equal to the marginal product of labor

$$\bar{w}_t = G^t F_N(G^t N, K_t) \qquad (3)$$

The wage rate of any young individual i differs from the average wage rate.[3]

$$w_{ti} = (1 + v_i)\bar{w}_t \qquad i = 1, \ldots, N \qquad (4)$$

where

$$1 + v_i \geq 0, \, i = 1, \ldots, N \quad \text{and} \quad \sum_{i=1}^{N} v_i = 0 \qquad (5)$$

---

3. Since the average real wage is equal to the marginal product of labor, the wage bill plus total returns to capital exhaust the product.

Individuals consume when both young and old. When young, they save part of their resources to consume during old age or to pass as a bequest to the next generation. Savings may be held as capital or as government bonds; bonds are perfect substitutes for capital in the portfolios of individuals, so both assets carry the same interest r. When old, individuals may get a bequest in the form of either government bonds or capital from their deceased parents. They use the bequest for consumption or to leave a bequest to their own children. All individuals have the same time invariant utility function[4]

$$u(c_t^1, c_{t+1}^2) + \beta V^{t+1}, \qquad 0 \leq \beta < 1 \tag{6}$$

where c_t^1 and c_{t+1}^2 are, respectively, first- and second-period consumption of a member of generation t; V^{t+1} is the maximum utility attained by the member's immediate offspring, and β measures the extent to which parents discount the utility of their children. The utility function is strictly concave, with positive and decreasing marginal utility of consumption in each period. Letting subscripts designate partial derivatives with respect to the subscripted variable,

$$u_i > 0, u_{ii} < 0 \qquad i = 1, 2 \quad \text{and } u \text{ is strictly concave} \tag{7}$$

Redistribution Through the Political Process

When old, each individual receives a lump-sum transfer of size S_t, which can be thought of as social security.[5] Total current expenditure, $S_t + (1 + r_{t-1})b_{t-1}$, is financed by a combination of lump-sum taxes on the current young and issuance of one-period government bonds that have to be repaid with interest in the next period. The government's budget constraint for period t is therefore[6]

$$P_t \equiv S_t + (1 + r_{t-1})b_{t-1} = T_t + b_t \tag{8}$$

where T is the lump-sum tax imposed on a young individual and b is the average quantity of one-period bonds per young or old individual. Each individual votes for the mix of financing that maximizes his utility. Social decisions about the intertemporal structure of taxation and redistribution are made by majority rule. A detailed description of the political process appears in Section III.

---

4. Consumption and V^{t+1} differ across individuals. We do not incorporate these individual differences in the notation explicitly. The constraint $\beta < 1$ reflects a positive time preference or some degree of selfishness or both on the part of parents. As pointed out by Buiter and Carmichael (1984) either factor alone is sufficient to deliver the constraint. (See also Burbidge, 1983, 1984; Carmichael, 1982.)

5. Since the focus of this chapter is on intergenerational redistribution, we abstract from the provision of public goods by government.

6. The budget constraint is normalized by the number of young (or old) individuals.

II. Who Is Likely to Be Bequest Constrained?

Characterization of Individual Economic Decisions for a Given Structure of Taxation and Redistribution

Each individual takes the structure of taxation and of social security benefits as given and chooses his consumption when young and old (c_t^1, c_{t+1}^2), the amount of his bequest, B_{t+1}, and the amount of resources, a_t, to be carried over from the first to the second period of life.[7] The first- and second-period budget constraints of a typical individual from generation t are, respectively,

$$w_t^N \equiv w_t - T_t = c_t^1 + a_t \tag{9a}$$

$$(1 + r_t)(B_t + a_t) + S_{t+1} = c_{t+1}^2 + B_{t+1} \tag{9b}$$

The choice of c_t^1, c_{t+1}^2, a_t, and B_{t+1} is determined by maximizing utility

$$V^t \equiv \max_{(c_t^1, c_{t+1}^2, a_t, B_{t+1})} [u(c_t^1, c_{t+1}^2) + \beta V^{t+1}] \tag{10}$$

subject to the two budget constraints in equation (9) and the additional constraint

$$B_{t+1} \geq 0 \tag{11}$$

The last constraint reflects the fact that there is no legal mechanism by which parents can borrow today against the future labor income of their children. They can leave their children nothing, but they cannot obligate their children's labor income.

Using equations (9a) and (9b) to eliminate a_t and c_{t+1}^2, substituting the resulting expression into (10), the Lagrangean for the problem in (10) is

$$\max_{(c_t^1, B_{t+1})} \{u[c_t^1, (1+r_t)w_t^N + S_{t+1} + (1+r_t)(B_t - c_t^1) - B_{t+1}] + \beta V^{t+1} + \lambda^t B_{t+1}\} \tag{12}$$

where $\lambda^t \geq 0$ is the Lagrange multiplier corresponding to the constraint B_{t+1}. Since the constraint qualification implies that either λ^t or B_{t+1} or both are equal to zero, the maximized value of the Lagrangean in (12) is equal to V^t from equation (10). Leading (12) by one period, substituting the resulting expression into (12), and continuing ad infinitum the problem in (12) may be expressed as a function of the decision variables of the current generation and of all future

---

7. B_{t+1} is the bequest left by generation t to generation $t+1$. It is set aside by generation t at the beginning of period $t+1$ and received with interest by the immediate offspring in period $t+2$ when they are old. It therefore carries interest r_{t+1} and the bequest received by generation $t+1$ is $(1 + r_{t+1})B_{t+1}$.

generations.[8] Recognizing that the decision variables of generation t directly affect only the utilities of generations t and $t+1$, and using the envelope theorem, we obtain the familiar first-order conditions for the problem of generation t as

$$u_1^t[\cdot] - (1 + r_t)u_2^t[\cdot] = 0 \tag{13a}$$

$$-u_2^t[\cdot] + \beta(1 + r_{t+1})u_2^{t+1}[\cdot] + \lambda^t = 0 \tag{13b}$$

$$\lambda^t B_{t+1} = 0 \qquad B_{t+1} \geqslant 0 \tag{13c}$$

Here subscripts designate partial derivatives with respect to the subscripted variables and the superscript designates the generation whose utility is being evaluated. For individuals who desire to leave a positive bequest, $B_{t+1} > 0$ and $\lambda^t = 0$. For these individuals equations (13a) and (13b) determine c_t^1 and B_{t+1}. For individuals who are bequest constrained, $B_{t+1} = 0$ and $\lambda^t > 0$; equations (13a) and (13b) determine c_t^1 and the shadow price of the bequest constraint, λ^t.

An intuitive understanding of λ^t is obtained by rewriting (13b) as $\lambda^t = u_2^t[\cdot] - \beta(1 + r_{t+1})u_2^{t+1}[\cdot]$. The first term on the right-hand side is the marginal utility of an additional unit of second-period consumption for generation t. The second term is the marginal contribution of the marginal utility of second-period consumption of the offspring to the parent's utility. When, at a zero bequest, the first term is larger than the second, the parent could have increased his utility by reallocating resources away from his offspring to himself. Since this is not possible under existing institutions, a positive wedge is created between the two marginal utilities. λ^t measures the loss of utility per unit of bequest that is created by the wedge resulting from the constraint $B_{t+1} \geqslant 0$.

An individual who is old in period t takes his first period consumption as given, since it was chosen in the previous period. Hence the only decision left to him is the allocation of his wealth and social security receipts between second-period consumption and his bequest. This leads to a first-order condition, as in (13b), and to a constraint qualification, as in (13c), both lagged by one period. Note that the old individual follows the same plan when old as he had planned for that period of life when he was young.

Who Is Bequest Constrained?

Individuals differ in their productivity and therefore in their real wage. They differ, also, in the real wage they expect their children, grandchildren, and later progeny to have. Since there is no uncertainty, each living individual knows the sequence of real wages that the future generations will earn. Differences in the real wages of parents also induce differences in their bequests to their children. Hence for given structures of taxation and redistribution, an individual's behavior as well as his welfare is fully determined by the bequest he receives, his wage, and the sequence of wage rates earned by all of his descendants. The

---

8. We assume that the present discounted value of the bequest at infinity is zero. Hence the individual's maximization problem is bounded. (See also Gale, 1983, Chap. 1, Sec. 8.)

purpose of this section is to show the influence of these known attributes on the size of the bequest that the individual desires to leave. This is done by considering an artificial problem in which the bequest constraint is removed (so that $B_{t+1}, j \geq 1$ can be either positive or negative) and by evaluating, using comparative statics, the effects of changes in individual attributes on the size of the desired bequest. The main results are summarized as follows:

Proposition 1: In the absence of bequest constraints in either the present or the future, and provided the utility function is strictly concave

(i) $$\frac{dB_{t+1}}{dB_t} = \frac{dB_{t+1}}{dw_t^N} > 0; \qquad \frac{dB_{t+1}}{dS_{t+1}} > 0$$

(ii) $$\frac{dB_{t+1}}{dw_{t+j}^N} < 0; \qquad \frac{dB_{t+1}}{dS_{t+j+1}} < 0 \quad \text{for all } j \geq 1$$

(iii) Provided individual wealth, $w_t^N + B_t$, is positive for all t,

$$\frac{dB_{t+1}}{dr_t} > 0 \quad \text{and} \quad \frac{dB_{t+1}}{dr_{t+j}} < 0 \quad \text{for all } j \geq 1$$

(iv) $$\frac{dB_{t+1}}{d\beta} > 0$$

Some of the results (parts (ii) and (iv) of the proposition) are due to Drazen (1978) and Weil (1987). All of the results are derived, within a unified framework, in Cukierman and Meltzer (1987a).

The proposition has intuitive appeal. Anything that makes the current generation richer—an increase in bequest received, an increase in the real wage rate, a decrease in current taxes, or an increase in social security payments—increases their bequest. Anything that makes any future generations richer decreases the bequest left by the current generation. Thus an increase in future wages, a decrease in future taxes, and an increase in future social security payments all tend to decrease the bequest chosen by the current generation.

Part (iii) of the proposition has a similar interpretation. Assuming that all individuals have positive wealth, an increase in the current interest rate makes an individual from the current generation richer, so he chooses a larger bequest. Conversely an increase in future interest rates makes the future generation richer, so the current generation reduces the current bequest. These results reflect the fact that the individual wants to spread the increase in consumption, made possible by an increase in r, over all periods. Part (iv) of the proposition states that the less the individual cares about the welfare of his offspring in comparison to his own welfare, the smaller the bequest he chooses to leave.

Taken as a whole, proposition 1 implies that for a given structure of taxation and redistribution the bequest chosen by an unconstrained individual depends on $w_t^N + B_t$ and on the sequence $w_{t+j}^N, j \geq 1$, of the net wage rates he expects

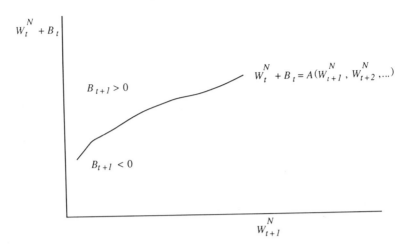

Figure 6-1.

succeeding generations of his family to earn. Let

$$w_t^N + B_t = A(w_{t+1}^N, w_{t+2}^N, \ldots) \tag{14}$$

be the value of $w_t^N + B_t$ that induces the current generation to pick $B_{t+1} = 0$, given the vector of future real wage rates. Parts (i) and (ii) of proposition 1 imply that the partial derivative of A with respect to w_{t+1}^N for $j \geqslant 1$ is positive. Figure 6-1 shows this relation. The upward-sloping curve represents the function $A[\cdot]$ for given values of w_{t+j}^N, $j \geqslant 2$. Parts (i) and (ii) of proposition 1 imply that individuals who are characterized by $(w_t^N + B_t, w_{t+1}^N)$ combinations that are above the curve choose a positive bequest. Those whose $(w_t^N + B_t, w_{t+1}^N)$ combinations lie below the curve leave a negative bequest.

An immediate implication of Figure 6-1 is that individuals whose wealth is low relative to the expected wage rate of their offspring, w_{t+1}^N, are likely to be bequest constrained since, in the absence of the constraint, they would have chosen a negative bequest. Conversely, individuals with relatively large wealth and relatively small values of w_{t+1}^N are more likely to be non-bequest-constrained. An increase in any of the wage rates w_{t+j}^N, $j \geqslant 2$, shifts, by part (ii) of proposition 1, the curve in Figure 6-1 upward so that the proportion of the population that is bequest constrained increases. It follows that the proportion of bequest-constrained individuals in the population is larger, the larger the rate of technological progress. Part (iii) of proposition 1 implies that an increase in the current rate of interest shifts the curve in Figure 6-1 downward, implying a decrease in the proportion of bequest-constrained individuals. An increase in any of the future real interest rates, or in the degree by which an individual prefers his own welfare to that of his immediate offspring, shifts the curve upward, increasing the proportion of bequest-constrained individuals in the population.[9]

III. The Political Process and Individual Preferences with Respect to the Intertemporal Structure of Taxation

Political decisions are made by majority rule. The young of generation t and the old of generation $t - 1$ vote to determine the size of the social security benefits to be paid to each member of generation t when he is old. This amount, denoted S_{t+1} (since it is paid out in period $t + 1$), is precommitted by social contract, which specifies that the voters' decision in t about S_{t+1} cannot be altered by those living in period $t + 1$. The voters in t also determine the allocation of financing of *current* government expenditure between taxes, T_t, and bonds b_t. When confronted with a choice between two different structures of taxation and social security, each individual votes for the schedule that maximizes his utility.

Since both S_t and b_{t-1} have been precommitted by the political decisions of the living in $t - 1$, the left-hand side of the government's budget constraint in (8) is predetermined from the point of view of those living in t. They are free to determine the structure of financing of P_t between current taxes and bonds but not the total that has to be financed. Their decisions about S_{t+1} and b_t, however, precommit those living in $t + 1$ to a total government budget of size P_{t+1}.

Reduction of the Political Choice Set

An individual who is young in period t is indifferent between different combinations of S_{t+1} and of b_t provided the future value, P_{t+1}, of those combinations is the same. The reason is that there is a perfect capital market between the first and second periods of life. An increase in S_{t+1} that is accompanied by a decrease in b_t reduces the disposable income of individuals currently young.[10] Their welfare is unaffected, since they can borrow against their higher social security at the market rate of interest to keep the present

---

9. Recently Altig and Davis (1987) have shown that when agents live for three periods, get most of their income in the middle period, and face borrowing constraints, the incentive of parents to bequeath is stronger than in the absence of such constraints. This structure encourages inter vivos transfers during a child's constrained period of life and narrows the set of parameters for which individuals are bequest constrained. But as long as there is a perfect capital market *within* an individual's lifetime (as assumed in the text), the basic thrust of proposition 1 carries through to a three-period model as well. Moreover, as long as there is a perfect capital market within an individual's lifetime, the timing of income, taxes, and bequests within the lifetime is irrelevant for individual welfare provided present values are preserved and only partial equilibrium effects are considered. However, the timing of those variables over the lifetime affects the general equilibrium value of the capital stock, and therefore welfare, through the consequent general equilibrium effects. For example, other things the same, the capital stock is lower with inter vivos bequests than with bequests that are transferred at death.

10. Since total expenditure in period t, P_t, is predetermined, a decrease in b_t implies through the government's budget constraint in (8) that the tax, T_t, on the young has to be increased.

value of their net wealth (including wages) intact. The same is true of an individual who is old in period t; the magnitudes S_{t+1} and b_t affect him only through the welfare of his immediate offspring who is young in period t. Hence all the individuals who vote in period t are indifferent between different combinations of S_{t+1} and b_t as long as they all have the same future value, P_{t+1}. It is therefore possible to set S_t arbitrarily at some fixed value—say S—and reduce the choice of P_{t+1} to that of choosing b_t for an arbitrarily given S.

Let

$$P_{t+1} = S + (1 + r_t)b_t \quad \text{for all } t \tag{15}$$

With S fixed, the choice of b_t uniquely determines P_{t+1} and, similarly, the choice of b_{t-1} uniquely determines P_t. In addition, from the government's budget constraint in (8), a choice of b_t also uniquely determines taxes, T_t, since P_t is predetermined by the political decisions of those who were alive in period $t - 1$. It follows that, given S, an individual's attitudes toward the triplet S_{t+1}, T_t, and b_t can be, equivalently, characterized by his attitude toward the one-dimensional variable b_t.

Individual Benefits from Intergenerational Reallocation of Taxes through Debt.

Barro (1974) demonstrated that when all generations are linked through an operative bequest motive, a reallocation of taxes across different generations via changes in the proportion of deficit finance does not affect individual welfare. The same is obviously true here for non-bequest-constrained individuals, as long as the present value of taxes is not altered and there is a perfect capital market.[11] However, individuals who are bequest constrained, or who expect that some future family members will be bequest constrained, benefit from substituting future for current taxes by means of bonds issued to finance the government budget. Since they are bequest constrained they would like to borrow against the future income of their children, but they are barred from doing so. Hence government bond financing increases the welfare of the bequest-constrained individuals even if the present value of all taxes is unchanged.

The current generation cannot choose which future generations will pay higher taxes when it votes to decrease current taxes and increase bond issues. Taxes are voted on each period. The next generation may decide to pass the tax increase to the following generation by voting to increase bond financing by an appropriate amount. The following generation, in turn, may either vote to postpone or to pay the tax, and so on.

The factors that determine whether there will be a majority for or against deficits or surpluses are discussed in Sections IV and V. Here it suffices to observe that, since expectations are rational and there is no uncertainty, those currently alive know which generation in the future will pay for a current tax

---

11. In fact such individuals are even indifferent to the time profile of P_t as long as the present value of this profile is the same.

reduction. The consequent change in welfare depends on the severity of the constraint, the number of future offspring that are also bequest constrained, and the time interval between the current generation and the generation to which taxes are ultimately shifted.

More formally it can be shown that a reduction in the taxes on generation $t + j$, financed by an increase in taxes on generation t, produces the following change in the welfare of generation t[12]:

$$\frac{dV^t}{dT_t} = -(1 + r_t)[\lambda^t + \sum_{s=1}^{j-1} \beta^s \prod_{i=1}^{s} (1 + r_{t+i})\lambda^{t+s}] \tag{16}$$

Equation (16) implies that if all individuals in the family between generation t and generation $t + j - 1$ are nonbequest constrained, all $\lambda^{t+s} = 0$ ($0 \leqslant s \leqslant j - 1$), and the intertemporal reallocation of taxes via bond financing does not affect the welfare of a young individual in period t. If at least one individual between generation t and generation $t + j - 1$ is bequest constrained, the shifting backward of taxes decreases the welfare of the currently alive individual, since at least one value of λ in equation (16) is positive.[13] The larger the number of bequest-constrained individuals between generation t and generation $t + j - 1$, and the tighter their constraints as measured by the values of λ, the larger the welfare loss caused by shifting taxes to the current generation.

IV. Crowding Out and the Determination of Government Debt by Majority Rule

The discussion of the previous section suggests that individuals in the economy may be classified into two broad groups. The non-bequest-constrained are indifferent to an intergenerational reallocation of taxes; the bequest constrained generally prefer to shift taxes to some of their descendants.[14] Each bequest-constrained individual favors an increase in bond financing and a corresponding decrease in current taxes that removes his constraint (and the constraint on all future generations of descendants who are similarly constrained). Once the

---

12. The derivation appears in Section IVB of Cukierman and Meltzer (1987a).

13. Note that the individual from generation t does not have to be bequest constrained with respect to his immediate offspring to dislike a backward shift of taxes.

14. Here we take the term *non-bequest-constrained* to mean that not only the currently alive individual but also all future generations of his descendants are non-bequest-constrained. The bequest constrained are generally composed of those who are constrained with respect to their immediate offspring (who may or may not be similarly bequest constrained) and those who desire to leave a positive bequest but know that some future generation will be bequest constrained. The bequest constrained like bond financing if it increases the tax burden of generations that come after the bequest-constrained individual. Otherwise they are indifferent to an intertemporal reallocation of taxes. For ease of exposition we focus in the text mostly on those who are immediately bequest constrained, but the same principles apply when one or more individuals in future generations is bequest constrained.

constraint is relaxed, the individual joins the group of non-bequest-constrained individuals and becomes indifferent to further reductions in taxes that are financed by issuing bonds.

If the political process produces some or all of the decrease in current taxes desired by the bequest constrained, disposable income of all currently alive young individuals increases. The reduction in taxes is matched exactly by an increase in bonds outstanding. The non-bequest-constrained individuals increase their bequest by the amount of the increase in their disposable income, as in Barro (1974). The bequest constrained use the increase in their disposable income to increase consumption. The resources to increase consumption are obtained by using part of the capital stock in the economy. Since bonds and capital are perfect substitutes in portfolios, the additional bonds are absorbed in the portfolios of the non-bequest-constrained individuals who release the physical resources needed to sustain the increased consumption of the bequest constrained.

The per capita increase in bonds is equal to the per capita decrease in taxes. A non-bequest-constrained individual increases his holdings of bonds, on average, by more than his proportional share of the increase in bonds, since the bequest-constrained individuals pay taxes but do not buy any bonds. Further, a non-bequest-constrained individual increases his bond portfolio, on average, by more than the increase in his disposable income. His excess purchase of bonds releases capital which is used by the bequest constrained to increase their lifetime consumption. The size of the decrease in the capital stock depends on the proportion of bequest-constrained individuals in the economy and on the degree to which each is bequest constrained. If no one is bequest constrained, capital does not decrease at all as a result of a given unit increase in bond financing; if some people are constrained, capital decreases by a fraction of the increase in bond financing. The proportion of bonds issued to capital displaced is therefore bounded between minus one and zero. The ratio is nearer to minus one the larger the proportion of bequest-constrained individuals in the economy. Formally, the crowding out ratio is

$$-1 < \frac{dk_t}{db_t} \leq 0 \qquad (17)$$

where k is the capital/labor ratio. Since the size of the population is fixed, the expression in (17) is identical to the crowding out ratio between the *total* amounts of bonds and of capital.[15]

The crowding out of capital changes the real rate of interest and the wage

---

15. The crowding out ratio exceeds -1 even in the extreme case in which everyone is bequest constrained prior to the shift in bond financing. The reason is that the aggregate value of planned bequests is zero, and the only capital that remains in the economy is held by young individuals who wish to transfer resources from the first to the second period of life. Since the marginal utility in both periods decreases with consumption, young individuals respond to the tax reduction by increasing consumption in both periods. As a result, the increase in the stock of bonds exceeds the decrease in the capital stock, and the crowding out ratio exceeds -1 even in this limiting case.

rates paid to individual workers. This creates two additional channels through which an increase in bond financing affects individual welfare. These general equilibrium effects of debt financing affect the welfare of both bequest-constrained and non-bequest-constrained individuals, since everyone's welfare is affected by changes in the interest rate and in individual wage rates.

The conbined effect of debt financing on welfare is the sum of the effects operating through three channels. First, an expansion in bond financing permits an intergenerational reallocation of resources within the family. Second, the marginal productivity of capital, and therefore the interest rate, changes. Third, the individual's wage rate changes. It is convenient, for expositional reasons, to examine first the political equilibrium when the last two general equilibrium effects do not operate. Formally this corresponds to the case in which the marginal productivities of capital and of labor do not depend on the capital/labor ratio. In this case a shift to bond financing affects individual welfare only through the intergenerational reallocation of resources that it makes possible.

Determination of the Debt in the Absence of General Equilibrium Effects

Each individual's decision about an increase in debt financing is, in this case, completely determined by whether or not he is bequest constrained. Each bequest-constrained individual prefers the amount of bond financing that frees him completely from his constraint. Hence those who face the most stringent bequest constraints favor the largest amount of debt. At the other extreme, the non-bequest-constrained are indifferent to the way governmental redistribution is financed. We model their indifference by assuming that they split their vote equally between different financing proposals. The bequest constrained always prefer financing proposals that include debt, so proposals to issue debt win in elections involving both groups.

Further, under the restriction that general equilibrium effects can be neglected, the majority favors a level of debt that is sufficiently large to free the bequest constraint of the most severely constrained individual. This choice frees all other bequest-constrained individuals from their constraints. The reasoning leading to this result is straightforward. The size of the government budget, P_t, is given. Suppose voters face two alternatives, one involving a positive quantity of bonds, b_{t0}, the other involving only taxes. Since the non-bequest-constrained split equally between the two proposals and all the bequest constrained vote for the proposal involving positive debt, a majority votes for the positive amount of bonds b_{t0}. Let $b_{t1} > b_{t0}$ be the level of debt at which a bequest-constrained voter is freed of his constraint when debt is b_{t0}. At b_{t0}, the non-bequest-constrained are indifferent to any level of debt greater than b_{t0}, so they split their votes equally for and against proposals calling for $b_{t1} > b_{t0}$. Hence there will be a majority in favor of any level of debt at which at least one individual is bequest constrained. Once the most severe bequest constraint is released, everyone is indifferent. Under complete certainty, there will be one vote each period to remove all bequest constraints.

Two qualifications are required. First, all constraints are released only if the capital stock is larger than the amount of additional consumption desired by all those who were bequest constrained.[16] Second, the amount of bonds cannot exceed the predetermined government outlays, P_t. If S in equation (15) is not sufficiently large, the constraint $b_t \leqslant P_t$ may prevent some of the most bequest-constrained individuals from reaching their preferred level of P_{t+1}. They vote to increase S_{t+1} in order to increase P_{t+1}.[17]

The main conclusion of this subsection is that in the absence of general equilibrium effects majority rule releases all bequest constraints by issuing debt, provided the existing capital stock is sufficiently large.[18] However, once the general equilibrium effects of a larger debt on factor returns are recognized explicitly, the non-bequest-constrained are no longer indifferent to the size of the debt. The following subsections analyze this more general case.

Individual Attitudes toward Debt in the Presence of General Equilibrium Effects

This section characterizes individual attitudes toward the level of the debt in the presence of general equilibrium effects. To find the total effect of an increase in debt financing on the welfare of different individuals in the economy, we first note that by using the recursive structure in (12) the two period's decision problem in this equation may be rewritten as the following infinite horizon problem:

$$\max_{(c_t^1, B_{t+1})} [u[c_t^1, (1 + r_t)w_t^N + S + (1 + r_t)(B_t - c_t^1) - B_{t+1}] + \lambda^t B_{t+1}$$

$$+ \beta \max_{(c_{t+1}^1, B_{t+2})} \{u[c_{t+1}^1, (1 + r_{t+1})w_{t+1}^N + S + (1 + r_{t+1})(B_{t+1} - c_{t+1}^1) - B_{t+2}]$$

$$+ \lambda^{t+1} B_{t+2}\} + \cdots) \qquad (18)$$

Differentiating (18) totally with respect to b_t, we obtain the total effect of a change in the level of debt on individual welfare. The total effect, shown in equation (19), has three components. The first, $dV^t/db_t|_{IR}$, is the the effect on current welfare from the intergenerational reallocation of resources induced by the change in

---

16. If the desired addition to consumption is larger than the capital stock, the solution is restricted by the existing resources.

17. All other individuals, being unconstrained, split evenly between any two proposals, so S_{t+1} rises to the point where all individuals become unconstrained. Hence $b_t \leqslant P_t$ does not change the result that the political process frees all individuals from their constraints. However, when the constraint $b_t \leqslant P_t$ is binding, part of the solution is found by increasing total outlays, P_{t+1}, for the next period, thereby increasing the likelihood that debt finance is higher next period. If only taxes, T_{t+1}, are raised to finance the increase in P_{t+1}, net wages of generation $t + 1$ decrease, pushing some of these individuals into the group of bequest-constrained individuals by part (i) of proposition 1. Since in each period the size of the debt is determined by the wishes of the most constrained individuals, the debt voted on in period $t + 1$ is larger.

18. In the absence of general equilibrium effects, a social planner would seek to remove the wedge also.

debt. Second is the effect on welfare from present and future changes in interest rates and wage rates. Third, the changes in interest rates and wages induced by the change in debt affect bequests; dB_{t+s+1}/db_t^T is the effect on bequests by current and future generations. Using the first-order condition from equation (13a), $u_1^t - (1 + r_t)u_2^t = 0$ for all t, the total derivative is

$$\frac{dV^t}{db_t} = \frac{dV^t}{db_t}\bigg|_{IR} + \frac{db_t^T}{db_t}\sum_{s=0}^{\infty}\beta^s u_2^{t+s}\left[(w_{t+s}^N + B_{t+s})\frac{dr_{t+s}}{db_t^T}\right.$$
$$\left. + (1 + r_{t+s})\frac{dw_{t+s}}{db_t^T} - \lambda^{t+s}\frac{dB_{t+s+1}}{db_t^T}\right] \quad (19)$$

where

$$b_t^T \equiv Nb_t \quad (20)$$

is the total amount of bonds in the economy. From equations (2), (3), (4), and (20) and the fact that the crowding out ratio measured using aggregates is equal to the ratio in per capita terms

$$\frac{dr_{t+s}}{db_t^T} = F_{KK}^{t+s}\frac{dK_{t+s}}{dK_t}\frac{dk_t}{db_t} = F_{KK}^{t+s}\frac{dk_t}{db_t} \quad (21a)$$

$$\frac{dw_{t+s}}{db_t^T} = (1 + v_{t+s})G^{t+s}F_{NK}^{t+s}\frac{dK_{t+s}}{dK_t}\frac{dk_t}{db_t}$$
$$= (1 + v_{t+s})G^{t+s}F_{NK}^{t+s}\frac{dk_t}{db_t} \quad (21b)$$

$$\frac{db_t^T}{db_t} = N \quad (21c)$$

where F_{KK}^{t+s} and F_{NK}^{t+s} are, respectively, the second partial derivative of the production function with respect to capital and the cross partial derivative of the production function between labor and capital in period $t + s$. We assume decreasing marginal productivity of capital and complementarity between labor and capital, so

$$F_{KK}^t < 0, \quad F_{NK}^t > 0 \quad \text{for all } t \quad (22)$$

v_{t+s} is a productivity class of generation $t + s$. Since $P_t = T_t + b_t$ and P_t is predetermined,

$$\frac{dT_t}{db_t} = -1 \quad (23)$$

Using equations (16), (21), and (23) in (19)

$$\frac{dV^t}{db_t} = (1 + r_t)\left[\lambda^t + \sum_{s=1}^{j-1} \beta^s \prod_{i=1}^{s}(1 + r_{t+i})\lambda^{t+s}\right] + N \sum_{s=0}^{\infty} \beta^s u_2^{t+s}[(w_{t+s}^N + B_{t+s})F_{KK}^{t+s}$$

$$+ (1 + r_{t+s})(1 + v_{t+s})G^{t+s}F_{NK}^{t+s}]\frac{dk_t}{db_t} - N \sum_{s=0}^{\infty} \beta^s u_2^{t+s}\lambda^{t+s}\frac{dB_{t+s+1}}{db_t^T} \quad (24)$$

Equation (24) is a general expression for the total change in welfare as a result of a one-unit increase in the deficit that is financed by an increase in taxes on generation $t + j$. The first term is the change in welfare due to the intergenerational reallocation of consumption. The second is the direct change in welfare due to the induced changes in factors returns. The third term is the change in welfare induced by the realignment in bequests due to the change in factor returns.

The Effects of Debt on the Welfare of a Non-Bequest-Constrained Individual

For a non-bequest-constrained individual, $\lambda^{t+s} = 0$ for all s. The only effects of debt issues (or withdrawals) are the induced effects on wages and interest rates arising from the change in the consumption of the bequest constrained. Equation (24) reduces to its second term,

$$\frac{dV^t}{db_t} = N\frac{dk_t}{db_t}\sum_{s=0}^{\infty} \beta^s u_2^{t+s}[(w_{t+s}^N + B_{t+s})F_{KK}^{t+s}$$

$$+ (1 + r_{t+s})(1 + v_{t+s})G^{t+s}F_{NK}^{t+s}] \quad (24a)$$

If there are no bequest-constrained individuals in the economy, the crowding out ratio dk_t/db_t equals zero, and deficits have no impact on welfare. This is not surprising since, in the absence of crowding out, factor returns are not affected by the way government expenditure is financed.

Equation (24a) implies that the increase in the interest rate and the decrease in the wage rate caused by the crowding out of capital have opposing effects on the individual's welfare. Since $w_{t+s}^N + B_{t+s} > 0$ for all $t + s$ and $dk_t/db_t < 0$, the increase in interest rates increases the individual's welfare while the decrease in real wages reduces his welfare. The net effect on welfare depends on the relative sizes of F_{KK} and F_{NK} and on the personal characteristics of the individual.

If the marginal product of capital is relatively sensitive to the quantity of capital (high $|F_{KK}|$) and wage rates are relatively insensitive to the quantity of capital (low F_{NK}), debt issues are more likely to increase than to decrease the individual's welfare. Conversely, when F_{NK} is high relatively to $|F_{KK}|$, the effect of debt issues and crowding out on welfare, through the decrease in wages, is more likely to dominate. If we assume that, as the capital/labor ratio decreases, F_{NK} increases relative to $|F_{KK}|$, the decrease in real wage rates ultimately dominates the rise in interest rates on the individual's welfare as the capital stock falls. Welfare declines, and a rising fraction of the non-bequest-constrained oppose further debt. Votes for surpluses, to increase the capital stock and wages, rise. A falling capital/labor ratio acts as a brake on the tendency to create deficits and ensures that deficits will disappear before the entire capital stock is consumed.[19]

For a given capital stock, different non-bequest-constrained individuals are affected differently by an increase in debt financing. Non-bequest-constrained individuals with relatively large bequests and relatively low labor productivity have large values of $w^N + B$ and low values of v. The increase in the interest rate dominates the change in their welfare, so they vote for more debt relative to current taxes. Conversely, non-bequest-constrained individuals who receive small (or zero) bequests and have relatively large labor productivity vote against a higher debt. The fall in wages dominates the change in their welfare.

We can summarize the effects of debt finance on the welfare and votes of the non-bequest-constrained in the following propositions. Those who have a relatively large fraction of nonhuman to human wealth favor a larger fraction of debt financing. Those who have a relatively large fraction of human wealth prefer a lower fraction of debt.[20] As the capital stock falls, some of the individuals who previously favored larger debts oppose further additions because reductions in wages become more important than increases in interest rates. This limits the vote for a larger debt.

The Effects of Debt on the Welfare of a Bequest-Constrained Individual

All of the welfare effects of changes in interest rates and wages carry over to bequest-constrained individuals. In addition, debt issues permit the bequest constrained to transfer resources from future generations to themselves. This is represented by the first term in equation (24); this term is always positive for bequest-constrained individuals.[21] The third term in equation (24) is the effect of changes in factor returns on the size of bequests. The marginal increase in debt and the induced changes in factor prices can remove the constraint from some who were previously bequest constrained. They become unconstrained. Others may find that the constraint remains binding but is less severe. For everyone

---

19. Obviously this result obtains even if F_{NK} increases relatively to $|F_{KK}|$ only after a sufficiently low capital stock.

20. Note that whether the ultimate effect of an increase in debt on welfare is positive or negative, it is stronger the larger the fraction of constrained individuals in the economy, since $|dk_t/db_t|$ is larger in this case. The existence of more bequest-constrained individuals magnifies the differential effects of debt on the welfare of non-bequest-constrained individuals with different structures of wealth.

21. A full discussion of the factors that determine this term is in part B of Section III and is not reproduced here. Recall, however, that the larger the number of bequest-constrained individuals among current and future generations, the larger are the benefits from intergenerational reallocation of resources.

who remains bequest constrained, $dB_{t+s+1}/db_t^T = 0$; the last term in (24) vanishes. For individuals who originally were near the margin, the product $\lambda^{t+s} dB_{t+s+1}/db_t^T$ may (but does not have to) differ from zero. By definition

$$\frac{dB_{t+s+1}}{db_t^T} = \frac{dB_{t+s+1}}{dw_{t+s}} \frac{dw_{t+s}}{db_t^T} + \frac{dB_{t+s+1}}{dr_{t+s}} \frac{dr_{t+s}}{db_t^T} + \frac{dB_{t+s+1}}{dB_{t+s}} \frac{dB_{t+s}}{db_t^T} \quad (25)$$

From proposition 1, dB_{t+s+1}/dw_{t+s}, dB_{t+s+1}/dr_{t+s}, and dB_{t+s+1}/dB_{t+s} are all positive. But dw_{t+s}/db_t^T and dr_{t+s}/db_t^T have opposite signs, so the sum of the first two expressions on the right-hand side of equation (25) is ambiguous. For a similar reason, the sign of dB_{t+s}/db_t^T is ambiguous as well. Moreover, since dB_{t+s+1}/db_t^T is non-zero only for individuals who are very near to being nonbequest constrained, it is multiplied by a value of λ^{t+s} which is close to zero. The expression for dB_{t+s+1}/db_t^T in (25) includes both positive and negative terms that tend to offset each other. It seems reasonable to assume that, even when there are some products in the last sum on the right-hand side of (24) that are non-zero, the terms in (25) do not dominate the sign of the expression for dV^t/db_t. Given this assumption, the change in welfare experienced by a bequest-constrained individual as a result of a one-unit increase in bond financing depends on three components: (1) benefits of intergenerational reallocation of resources, (2) the increase in welfare from a higher return on assets, and (3) the decrease in welfare due to the decrease in wage rates.

Characterization of the Voting Equilibrium

We turn now to a more precise characterization of the political equilibrium in the presence of general equilibrium effects. To avoid potential problems of cycling with majority rule we assume that V^t is a concave function of b_t.[22] Under this condition, V^t is a single peaked function of b_t and the level of b_t for which

$$\frac{dV^t}{db_t}[b_t; B_t, (w_{t+j}, j = 0, 1, 2, \ldots)] = 0 \quad (26)$$

is the stock of debt most preferred by the individual under consideration. Since individuals differ in bequests received, in their wage rates, and in the sequence of

---

22. A more detailed discussion of some underlying conditions for this concavity appears in Appendix C of Cukierman and Meltzer (1987a). We assume, for simplicity, that (although they can perfectly predict future political outcomes) the currently alive voters do not take into consideration the effect of their choice of b_t on the votes of future generations. In other words, they do not vote strategically. Instead, as in a regular Nash equilibrium, they take those votes as given. But they do take into consideration, when voting, the general equilibrium effects of current political outcomes on the welfare of all their offspring. Note that having atomistic agents take into account general equilibrium ramifications of their actions is, in this case, rational behavior on their part, since they can affect returns to factors in their possession through voting.

wage rates they expect for future generations, the most preferred value of b_t differs among individuals. Let

$$b_t^* = b^*[B_t, (w_{t+j}, j = 0, 1, 2, \ldots)]$$

be the value of b_t most preferred by an individual who received a bequest of size B_t, has a wage rate w_t, and expects a sequence of wages $(w_{t+j}, j \geq 1)$ for future generations of descendants. The value of b_t is restricted from above by the predetermined value of P_t. We assume government debt cannot be negative,[23] so

$$0 \leq b_t \leq P_t \tag{27}$$

Hence

$$b_t^* = b_t \quad \text{from equation (26) if } 0 \leq b_t \leq P_t$$

$$b_t^* = 0 \quad \text{if } \frac{dV^t}{db_t}[0; B_t, (\cdot)] < 0 \tag{28}$$

$$b_t^* = P_t \quad \text{if } \frac{dV^t}{db_t}[P_t; B_t, (\cdot)] > 0$$

Since the utility of each individual is a single peaked function of b_t, there exists a unique median range for b_t that will defeat any other value of b_t outside this range when voting is by majority rule.[24] Let b_{td} be a point in the median range. To avoid unnecessary complications, we assume that the arbitrary value of S is set at a level at which some taxes are paid; this ensures that the maximum value of b_{td} is smaller than P_t. Hence, even if there are individuals who would like to increase the level of S, they never have a majority. Consequently, S does not change over time, as assumed in Section III. The precise location of the median depends on the relative frequencies of the three main types of individuals described in the previous subsection.

---

23. When government is allowed to set taxes above the level necessary to finance P_t in order to lend to the public, the outstanding government debt may be negative. In this case the constraint $b_t \geq 0$ in (27) and the second line of (28) are no longer relevant. Instead the first condition in equation (28) holds for the entire range $b_t \leq P_t$. Assuming government lends equally to all young individuals, the non-bequest-constrained are indifferent to the existence of this additional option. The higher taxes and loans mean that they have fewer resources in the second period of life, and some of their offspring have a lower tax burden. They compensate for that by an appropriate downward adjustment in their bequests. The bequest constrained must reduce their consumption, since the loan is for one period only. As a result the capital stock increases. Such a policy is likely to be favored by non-bequest-constrained individuals with a relatively large ratio of human to nonhuman wealth.

24. If the distributions of B_t and of (w_{t+j}) across descendants is sufficiently dense, this range is quite narrow. In the limit when the distributions of B_t and of (w_{t+j}) are continuous, the median b_t reduces to a single point.

V. Economy-wide Conditions that Are Conducive to Debt and Deficits

Who Votes For and Against a Large Debt?

We saw that non-bequest-constrained individuals may oppose debt if a large fraction of their wealth is human wealth. Bequest-constrained individuals may oppose debt also, if their loss from the decrease in wage rates is larger than the sum of their gains from the increase in interest rates and the increase in the availability of current resources. Inspection of equation (24) and proposition 1 suggests that the latter group is likely to include individuals whose total wealth is modest but who have a relatively large fraction of wealth in human capital. They are bequest constrained, but their welfare loss because of the constraint is relatively small. Anything that makes the shadow prices λ^{t+s} of the bequest constraint not too large, such as an expectation that the wages of future generations will be only modestly higher than the wage of the current generation, increases the likelihood that a bequest-constrained individual will vote against a large debt.

Bequest-constrained individuals who are likely to vote for higher debt include individuals with low total values of $w_t^N + B_t$—small inheritance and low productivity—who expect their offspring to have productivities and wage rates substantially higher than their own. Such individuals choose to increase their own consumption at the expense of their descendants. They suffer a larger loss from being bequest constrained than from the induced decrease in their real wage rate when the constraint is relaxed. They are therefore likely to vote for a larger debt.

Paradoxically, individuals with large $w_t^N + B_t$, particularly if it is composed of a large component of inherited wealth, also are likely to vote for large debt. By proposition 1 such individuals are unlikely to be bequest constrained, so they derive no benefits from an intergenerational reallocation of consumption. However, their attitude toward a larger debt is likely to be dominated by the induced increase in interest rates which increases the major component of their income.

To summarize, the coalition favoring increased debt and deficits includes several different groups. Individuals with large and small inheritance will be in the coalition. If those with large inheritance are "rich" and those with small, or zero, inheritance are "poor," some rich and some poor favor deficits. Their reasons differ, however. The poor vote for deficits to transfer resources from future generations in their family to themselves. The rich vote for deficits to increase the return on their portfolios, particularly if investment income is a large part of their total income. Many of the voters in between, the middle class, are likely to oppose deficits. This is particularly true of any voter with relatively high productivity. Typically, such voters are not bequest constrained, or the constraint is not severe. The welfare loss from wage reduction is likely to be larger than the welfare gain from redistribution and from higher returns to capital.

The coalitions favoring and opposing deficits shift as the numbers in the various groups change. Changes in current and expected wages and interest rates, reflecting changes in current and prospective productivity of labor and capital, induce changes in the direct and indirect effects of debt on the welfare of individual voters.

Economy-wide Conditions Conducive to Larger Debts

The larger the rate of labor augmenting technological progress, the larger is the fraction of individuals who are bequest constrained and the more they stand to benefit from an increase in the amount of debt financing. Hence debt financing increases with the rate of technological progress.

The larger the fraction of individuals with a relatively small total wealth (both human and nonhuman), the larger the fraction of individuals for whom loosening of the bequest constraint is a prime consideration, and the larger, therefore, the level of debt preferred by the median voter.[25]

The smaller the fraction of individuals whose main source of income is wages, the smaller the fraction of individuals who oppose debt because of its downward effect on wages, and the larger the level of the debt. We saw that the coalition favoring larger debts is composed of individuals with extreme values of wealth and income. Hence larger debts are more likely the more spread out is the distribution of individuals by total wealth or income.

The more sensitive is the return to capital to a change in the capital/labor ratio, the stronger the upward effect of an increase in debt on the return to capital and the larger the level of debt preferred by the median voter. The less sensitive the level of wages to a change in the capital/labor ratio, the smaller the downward effect of a larger debt on wage rates and the larger the debt level picked by the political process. Higher expected longevity (that results in longer time spent in retirement) is also conducive to higher debt, since it increases the utility of the older individual's own consumption and makes it more likely that he is bequest constrained and that he prefers, therefore, a larger debt. This element can be modeled formally by using the marginal utility of consumption in the second period of life as a proxy for the length of time spent in retirement. A detailed analysis appears in Cukierman (1986).

The discussion of this subsection is summarized in the following proposition.

Proposition 2: Under majority rule a larger debt is more likely

a. the larger the expected rate of growth of the economy
b. the larger the fraction of individuals below a certain level of income and wealth
c. the smaller the fraction of individuals whose main source of income is wages
d. the more spread out the distribution of individuals by total wealth or income

---

25. Here we assume that the distribution of wage rates and bequests received is sufficiently dense so that the median values of b_d can be approximated by single points.

e. the more sensitive the return to capital to a change in the capital/labor ratio
f. the less sensitive the level of wages to a change in the capital/labor ratio
g. the higher the expected longevity

Economy-wide Conditions Conducive to Deficits

By definition, deficits are created when the national debt increases. Hence any of the factors in proposition 2 that increase debt also increase the deficit for a time. More precisely, a deficit is created when the level of debt preferred by the decisive voter in period t, b_{dt}, is larger than the level of debt preferred by the decisive voter of the previous period, $b_{d,t-1}$. Thus an increase in the rate of growth between periods t and $t+1$ in comparison to the rate of growth between periods $t-1$ and t is likely to produce a deficit in period t. The reason is that, in comparison with the previous period, more individuals are bequest constrained and the constraints are more severe. The number of individuals who favor a relatively large b_t increases.

Deficits are also more likely when the capital stock increases. With a rising capital stock, the negative effect of higher debt on wages becomes less important for welfare compared with the positive effect of total returns from assets.[26] As a result, the ideal level of debt increases, particularly among the non-bequest-constrained. The increase tends to make b_{dt} larger than $b_{d,t-1}$.

The higher the total budget, P_t, that has to be financed, the higher is b_{dt}. If an increase in the budget is financed only by current taxes, net current wages decrease. As shown in part (i) of proposition (1), more individuals move into the ranks of the bequest constrained, and the ideal values of b_t among bequest-constrained individuals rises. As a consequence, b_{dt} increases. If P_t is high mainly because social security benefits are high, it is likely that b_{dt} is also larger than b_{t-1}, so there is a deficit.

Given that the deficit equals the first difference of the debt, the implications of proposition 2 for deficits are summarized in proposition 3.

Proposition 3: Budgetary deficits are larger under majority rule in periods in which *there has been an increase in*

a. the expected rate of growth of the economy
b. the fraction of individuals below a certain level of income and wealth
c. the fraction of individuals whose main source of income is not from wages (rentiers)
d. the spread of the distribution of income
e. expected longevity

Preliminary evidence on changes in the functional and size distribution of income as well as in longevity in the United States between the 1970s and the

---

26. This follows from the assumption that the ratio $|F_{KK}|/F_{NK}$ increases when the capital/labor ratio increases.

1980s supports parts c and e of the proposition and does not contradict the other parts (Cukierman & Meltzer, 1987b).

VI. Concluding Remarks

This chapter has presented an integrated economic and political theory of public debt determination that is based on redistributional considerations (across and within generations) in the presence of differences in abilities and wealth. A basic implication of the theory is that the existence of a positive national debt is directly traceable to the existence of a sufficient number of individuals who desire to leave negative bequests but are prohibited from doing so. By voting for deficits, they increase their consumption, crowding out capital but reducing the severity of the bequest constraint.[27]

Although our model has many of the features found in Barro (1974), we reach a very different conclusion. Debt issues have macroeconomic effects on interest rates, wages, and the stock of capital even when the present value of future taxes equals the value of the debt. The differences are the result of redistribution, where there are some bequest-constrained individuals, features neglected in Barro's model.[28]

Deficits are often incurred during wars. Our chapter focuses on redistribution and abstracts from wartime defense and other public goods. Some of the results extend, however, to the case of an exogenously given, but possibly fluctuating, level of expenditures on public goods. In particular, the theory implies that an increase in government expenditures (possibly due to war) induces, as in Barro (1979), an increase in the public debt even with nondistortionary lump-sum taxes. The mechanism that produces this result differs from the one suggested by Barro. If all the increase in government expenditure is financed by higher taxes, wages net of taxes fall absolutely and relative to future (postwar) generations. The decline in net wages increases the fraction of individuals who are bequest constrained and who vote for a larger debt. As a result, the level of debt most preferred by the decisive voter increases, inducing the political process to use both taxes and debt to finance the increased level of public good expenditures. More generally, the model implies that deficit financing is more likely when government expenditures increase, and surpluses are more likely when expenditures decline.

To concentrate on the type of effective redistribution ignored by Barro, we have neglected relevant factors affecting taxes, spending, bequests, and consumption. To separate the implications of redistribution from the effects of tax distortions for the level of the national debt, we have assumed lump-sum taxes.

27. In practice they would be joined by people without offspring who increase their consumption by taxing the offspring of others.

28. It is interesting to note in this context that some types of diversity imply the existence of at least one bequest-constrained individual in general equilibrium. Thus Aiyagari (1989) shows within the context of a pure exchange economy that when individuals have different rates of time preference there is at least one bequest-constrained individual in the economy. This is the individual with the highest rate of time preference.

Obviously, tax-induced distortions are important in practice. The principles developed here provide some insights that may be relevant for the analysis of tax-induced distortions. Those who would benefit from transfers financed by distortionary taxes on current and future generations probably include many of the same people who benefit from debt finance. Those with low wages, relative to the wages expected by the next generation, would try to tax future wealth. Those who gain from crowding out capital by issuing debt would be willing to crowd out capital by taxing returns to capital.

Feldstein (1976) suggests that a significant part of the transfer of resources from parents to children takes place when both generations are still alive. These transfers include both consumption and investment, particularly the purchase of education and other investments in the human capital of the children. Investments in education affect the productivity of offspring and their expected future wages. The traditional overlapping-generations framework, used here, does not incorporate this element explicitly. However, we can interpret the first-period consumption of an individual as including expenditures on education. On this interpretation, consumption in the first period of life yields higher utility and is therefore higher. Reallocation of spending to the first period makes it more likely that the decisive voter is bequest constrained and increases the severity of the constraint. The decisive voter, therefore, prefers a larger national debt.

Our focus has been on individual preference for positive debt induced by bequest constraints. Drazen (1978) points out that if the rate of return on investment in human capital is higher than the return on physical capital, parents prefer to invest at least some resources in the human capital of their children not only for the sake of their children but also to provide for their own retirement. If the bequest they wish to leave at death is larger than the level of investment in human capital at which the returns on human and physical capital are equal, they are indifferent between taxes and debt. However, if the reverse is true, they prefer some debt financing to capture back at retirement the excess of investment in the human capital of their children over their desired bequest, while still allowing the children to enjoy the higher returns from education. In this case (abstracting from general equilibrium effects), individuals have a strict preference for debt whenever their desired bequest is lower than the level of investment in human capital at which the return on this investment is equal to the return on physical capital. In contrast, such a strict preference develops here at the point at which desired bequests become negative. Obviously incorporation of this additional element raises the level of debt most preferred by the decisive voter and therefore the level of the national debt. Consequently, it implies that deficits are more likely to occur during periods in which the return on human capital rises relative to the return on physical capital.

References

Aiyagari, R. S. (1989). "Equilibrium Existence in an Overlapping Generations Model with Altruistic Preferences." *Journal of Economic Theory*, *47*, (Febr.), pp. 130–52.

Altig, D. and Davis, S. J. (1989). "Government Debt, Redistributive Fiscal Policies, and the Interaction between Borrowing Constraints and Intergenerational Altruism." *Journal of Monetary Economics, 24* (July), pp. 3–29.

Barro, R. J. (1974). "Are Government Bonds Net Wealth?" *Journal of Political Economy, 82* (Nov./Dec.), pp. 1095–1117.

Barro, R. J. (1979). "On the Determination of the Public Debt." *Journal of Political Economy, 87* (Oct.), pp. 940–71.

Buiter, W. H. and Carmichael, J. (1984). "Government Debt: Comment." *American Economic Review, 74* (Sept.), pp. 762–65.

Burbidge, J. B. (1983). "Government Debt in an Overlapping-Generations Model with Bequests and Gifts." *American Economic Review, 73* (Mar.), pp. 222–27.

Burbidge, J. B. (1984). "Government Debt: Reply." *American Economic Review, 74* (Sept.), pp. 766–67.

Carmichael, J. (1982). "On Barro's Theorem of Debt Neutrality: The Irrelevance of Net Wealth." *American Economic Review, 72* (Mar.), pp. 202–13.

Cukierman, A. (1986). "Uncertain Lifetimes and the Ricardian Equivalence Proposition." W. P. No. 45–86, The Foerder Institute for Economic Research, Tel-Aviv University (Dec.).

Cukierman, A. and Meltzer, A. H. (1987a). "A Political Theory of Government Debt and Deficits in a Neo Ricardian Framework—Extended Version." Unpublished Manuscript, July 1987. Pittsburgh: Carnegie Mellon University, Graduate School of Industrial Admin.

Cukierman, A. and Meltzer, A. H. (1987b). "A Political Theory of Government Debt." Presented at the December meeting of the American Economic Association, Chicago.

Diamond, P. A. (1965). "National Debt in a Neoclassical Growth Model." *American Economic Review, 55* (Dec.), pp. 1126–50.

Drazen, A. (1978). "Government Debt, Human Capital, and Bequests in a Life-Cycle Model." *Journal of Political Economy, 86* (June), pp. 505–16.

Feldstein, M. S. (1976). "Perceived Wealth in Bonds and Social Security: A Comment." *Journal of Political Economy, 84* (Apr.), pp. 331–36.

Gale, D. (1983). *Money: In Disequilibrium*, Cambridge, Mass: Cambridge University Press.

Meltzer, A. H. and Richard, S. F. (1981). "A Rational Theory of the Size of Government." *Journal of Political Economy, 89* (Oct.), pp. 914–27.

Roberts, K. W. S. (1977). "Voting over Income Tax Schedules." *Journal of Public Economics, 8* (Dec.), pp. 329–40.

Romer, T. (1975). "Individual Welfare, Majority Voting, and the Properties of a Linear Income Tax." *Journal of Public Economics, 4* (Feb.), pp. 163–85.

Samuelson, P. A. (1958). "An Exact Consumption Loan Model of Interest with or without the Social Contrivance of Money." *Journal of Political Economy, 66* (Dec.), pp. 467–82.

Weil, P. (1987). "Love Thy Children: Reflections on the Barro Debt Neutrality Theorem." *Journal of Monetary Economics, 19* (May), pp. 377–91.

7

A Positive Theory of Discretionary Policy, the Cost of Democratic Government, and the Benefits of a Constitution

ALEX CUKIERMAN AND ALLAN H. MELTZER

I. Introduction

A main implication of recent developments in economic theory is that governments can increase welfare by using rules, usually state contingent rules, instead of discretion. The bases for this conclusion range from the formal demonstration of dynamic inconsistency, introduced by Kydland and Prescott (1977), to the more general argument that discretion increases the public's uncertainty.

While the issue of rules versus discretion is far from closed in economic theory, recent work in the rational expectation tradition has strengthened the case for rules. The case for rules has attracted few practitioners, however. Governments maintain discretionary policies in many areas and resist efforts to adopt monetary and fiscal rules or the fixed tax and subsidy rules to control pollution advocated by many economists. Policymakers appear to prefer discretion to rules even when arguments in favor of precommitments seem compelling. A major purpose of this chapter is to explain this phenomenon from a positive point of view and to investigate the welfare implications of retaining discretion.

The economic policies chosen by governments depend on the aims or goals of policymakers and the constraints under which they operate. Here we take the view that public officials choose the economic policies that are most likely to get them reelected. Like the entrepreneur of economics who strives to maximize

---

We are indebted to Motty Perry for a perceptive criticism of an earlier draft. We also would like to thank colleagues at Carnegie Mellon University, Richard J. Sweeney and a referee for useful comments on a previous draft.

profits, the politician acts to maximize the likelihood of being reelected.[1] Unlike the entrepreneur, however, politicians must maximize support at a particular point in time—when elections are held—so that they have an incentive to choose policies that are acceptable to the public when they vote.

In the presence of unanimity about social goals and symmetric information, political competition within a democracy is likely to lead to socially optimal economic policies. Where the public has only imperfect information about the actions of government and the state of the economy, the socially optimal economic policies and the support for maximizing choices may diverge. This chapter shows that in the presence of uncertainty about the future optimal settings of policy instruments, governments will seek to retain flexibility in the choice of instruments.[2] Flexibility is achieved in one of two ways. Either policymakers have discretion, or they are committed in advance to follow a socially optimal, contingent, decision rule. When the policymaker's objective is to maximize social welfare, the result of both arrangements is identical. However, when the policymaker is a politician who strives to maximize support on election day, discretion leads to a socially suboptimal outcome. In what follows we elaborate on the origin of this result.

Governments that have discretionary authority differ in their ability to interpret events and forecast the future. Governments with better forecasting ability are more likely to produce greater welfare, so they are preferred by voters. Since the public has incomplete information on the forecasting ability of an incumbent government, rational voters use the level of welfare experienced under this government as an indicator of its forecasting ability and future performance.[3] As a result, incumbents' reelections depend on welfare generated during their term of office. The incumbent acts to increase welfare by the end of his term even if the policy involves a substantial and above-optimal loss of welfare after the election.[4] The welfare loss is directly traceable to the existence of periodic elections, so we call it "the cost of democracy." It is larger the greater the frequency of elections.

Recent literature in political economy provides evidence that (1) current and past economic conditions affect the popularity of governments and (2) at times governments choose economic policies to increase public support at election

---

1. This view of the "political entrepreneur" was forcefully developed by Downs (1957). This is not to deny the possible influence of ideologies on policy making. Since ideologies seem less important for the understanding of the preference for discretion, however, we abstract from them. In a later section we separate individuals according to their preference for activism.

2. This preference for flexibility is analogous to the demand for flexibility by private entrepreneurs facing an irreversible investment decision in an uncertain world. See Cukierman (1980) and Bernanke (1983) for examples.

3. The use of imperfect signals under partial information is the basis of well-known work on the business cycle. See Lucas (1973, 1975).

4. There are many examples that fit this paradigm, but an extreme case is the policy of the Israeli government prior to the 1981 elections. Despite a large balance of payments deficit, the government maintained the rate of exchange at a low level to keep import prices low. They borrowed foreign exchange and substantially increased international debt. The public reelected the government. Later on, foreign exchange controls and other austerity measures had to be imposed.

time.[5] This chapter reconciles this evidence with the view that a rational public should look forward rather than backward when voting. When there are limitations on available information and persistence in the attributes of different governments, the public rationally uses past performance as a signal about future performance.

A constitutional commitment to a socially optimal contingent choice of policy instruments could eliminate the cost of democracy without losing the flexibility needed for the maximization of social welfare. However, a constitution raises serious enforcement problems. Without full transmission of information by government to the agency that monitors and enforces compliance with the constitution, government does not fully bear the cost to the public of discretionary policy. As a consequence, even when a socially optimal constitution is enacted, government is often tempted not to abide by it.

In the presence of a voting public with diverse objectives, the government's choice of policy instruments determines both total welfare and its distribution across different groups. A government seeking reelection sets its policy instruments to achieve the distribution of welfare that maximizes its reelection prospects. The government behaves as if it maximizes the welfare of the mean voter during its period in office.[6] We show that the existence of diversity in the electorate does not eliminate the cost of democracy. The reason is that policy actions that increase the prospects for reelection move the economy away from the Pareto efficient frontier and reduce welfare.

The structure of the model used to illustrate these ideas is presented in Section II. The choice of instruments by an apolitical social planner is also presented, as a benchmark, in this section. We show that, in the presence of uncertainty, flexibility in the choice of instruments is necessary for a social optimum. Section III derives the choice of policy instruments by a politically motivated government facing imperfectly informed voters and shows that discretionary policy leads to suboptimal choices of policy instruments. The reason for this result is amplified in Section IV, where we find that the cost of democracy disappears when the public has the same information as the government. Section V develops the social benefits of a constitution and shows that due to asymmetric information between government and the public, the government is unlikely to adhere strictly to the constitution. An illustration that maps into the general framework of the previous sections is discussed in Section VI. It concerns the amount of resources drafted into the production of a public good. Section VII generalizes the analysis to the case in which various groups in the population differ in the degree of activism they prefer. Some concluding comments complete the chapter.

---

5. For evidence on the effect of economic conditions on the popularity of incumbents, see for example Frey and Schneider (1978) and Fischer and Huizinga (1982). Papers on the political business cycle by Nordhaus (1975), MacRae (1977), and others suggest that the choice of economic policy is not independent of the election cycle. See, however, McCallum (1978) and Meltzer and Vellrath (1976). No clear conclusion has emerged from this literature.

6. The mean voter plays the same pivotal role here as the median voter in Meltzer and Richard (1981).

II. The Social Planner's Problem

In the economy we consider, social welfare depends on the realizations of a random state variable and on the settings of a policy instrument that is chosen by government. Let x_t and a_t be, respectively, the realization of the state variable and the setting of the policy instrument in period t. The state variable x_t represents events that are beyond the control of either the private sector or the government. Examples of these events are unpredictable changes caused by nature or by other countries.[7]

Government is chosen in a democratic election for an office term of n periods. Elections are held at the end of each office term. The government's main objective is to be reelected. The public is concerned about its welfare and rewards a government for its performance. An incumbent government is more likely to be reelected the higher the level of social welfare during its term of office.[8]

Policies chosen today affect the economy's performance in the current and immediately following periods, so welfare in the current period depends on instrument settings in the past and current period. Formally, social welfare in period t is inversely related to the loss function,

$$L_t = (a_{t-1} - x_t)^2 + (a_t - x_t)^2 \tag{1}$$

Here x_t is a random normal variate with a zero expected value and variance σ_x^2. Losses increase nonlinearly when $a \neq x$. The fact that past policies affect both past and present welfare is a crucial element of our analysis.

The realization of x_t does not become known to government until the end of period t. The beginning of period t is the latest time at which the policy instrument, a_t, can be set to affect behavior in t. The government can, if it wishes, set a_t at an earlier time by precommitting to a particular path for the policy instruments. By waiting, the government obtains additional information. Specifically, at the beginning of period t the government obtains noisy indicators for x_t and x_{t+1} in the form of observations on the variables

$$y_t^0 = x_t + \varepsilon_t^0 \quad \text{and} \quad y_t^1 = x_{t+1} + \varepsilon_t^1 \tag{2}$$

where ε_t^0 and ε_t^1 are normally distributed white noise processes with zero mean and variance σ_ε^2, and are statistically independent of x and of each other. Section VI discusses a specific illustration of governmental actions and random state variables that affect social welfare and that map into the general framework presented here.

---

7. Changes in the parameters of behavioral equations and changes in technology are additional sources of unpredictable changes but are not considered in our formal model.

8. The evidence supporting such a relationship is surveyed in the context of public attitudes toward inflation and unemployment in Schneider and Frey (1984). See also footnote 5 in this chapter. Section III shows that this voting pattern is rational if the public is less than fully informed.

The government has no incentive to give accurate information about y_t to the public. In fact, complete revelation precludes the use of economic policy to improve election prospects. Even if the public has a noisy indicator of its own, the public's indicator is an imperfect substitute for the government's information. The reason is that the public does not know the government's forecasting ability, so it cannot separate fully the effects of government policy from other forces affecting x_t or be certain about the information that the government had when it chose policy actions.[9] In particular, one period before the election the public has incomplete information on the states of nature, x_t, realized during the term of the incumbent government and the instruments, a_t, chosen by the government. The public experiences changes in welfare, so it knows the level of welfare, L_t, experienced in each period. The public also knows the variance of states of nature, σ_x^2, but it cannot determine how much of the welfare level is due to nature and how much to either current or past governmental actions. The best it can do is draw inferences about the relative contributions of nature, or chance, and policy to its welfare.

Governments make forecasts as part of the policy-making process. Governments differ from each other in their ability to make precise forecasts of future states of nature, and for this reason policies differ. Each government is characterized by a different value of the noise variance σ_ε^2. This variance is unknown to the public, but the public makes inferences about σ_ε^2 from the level of welfare experienced during the incumbency period.

To focus on the main issue of the chapter with the fewest complications, we assume that the social welfare function is linear in the sum of the L_t. The public is risk neutral and does not have time preference, as in (3). (The qualitative results of the analysis are unaffected by the degree of time preference.) Differences between a and x impose costs that are nonlinear, as in (4). Maximization of social welfare by an apolitical benevolent government or a social planner who is in office from period 1 through period n involves the following problem:

$$\min_{(a_1,\ldots,a_n)} E_g \sum_{t=1}^n V_t \qquad (3)$$

where

$$V_t \equiv (a_t - x_t)^2 + (a_t - x_{t+1})^2 \qquad (4)$$

and E_g is the expected value operator conditioned on the information available to government. The government must choose a value of the policy instrument, a_t, before the values of all the state variables are known with certainty.[10] We

---

9. The critical point is that there is asymmetry in information available to the public and the government. Canzoneri (1985) uses an assumption similar to ours. In his model, the Federal Reserve's forecast of money demand is private information.

10. Equation (1) shows that current welfare depends on current and past policy actions and the current state of nature. Equation (4) shows the contributions to current and future welfare that are affected by the current instrument setting. The latter is more convenient for solution of equation (3).

restrict attention to the period over which the choice of instruments by a one-term government affects the public's welfare.[11] All other periods are irrelevant for the decisions to be made by this government.

Although our main interest is the behavior of a government that maximizes the likelihood of being reelected, we require a benchmark or standard to evaluate government action. We use as our standard a social planner who is concerned only with social welfare. A necessary condition for the minimization of the expected value of social losses (maximization of social gains) in (3) is

$$\min_{(a_t)} E_g V_t \quad t = 1, \ldots, n \tag{5}$$

We consider, first, the effect on social welfare of a decision to set the value of a_t before the arrival of period t. To evaluate this policy of precommitment, we compare the minimized value of the objective function in (5) for three alternative cases: (i) a_t is precommitted in period $t - j$ where $j \geq 2$; (ii) a_t is precommitted in period $t - 1$; (iii) a_t is chosen only in period t after y_t^0 and y_t^1 are revealed to the government. The objective function in (5) specializes in each of these cases:

$$\min_{\{a_t\}} E_{g,t-j} V_t \quad j \geq 2 \tag{6a}$$

$$\min_{\{a_t\}} E_{g,t-1} V_t \tag{6b}$$

$$\min_{\{a_t\}} E_{g,t} V_t \tag{6c}$$

In (6), the second subscript on the expectation operator denotes the information set available to the social planner when choosing a_t. For example $E_{g,t} V_t$ denotes the expected value conditioned on information available to government at the beginning of period t. The optimal values of the policy instruments for each of the problems in (6) are, respectively,

$$a_t = \tfrac{1}{2} E_{g,t-j}(x_t + x_{t+1}) = 0 \quad \text{for } j \geq 2 \tag{7a}$$

$$a_t = \tfrac{1}{2} E_{g,t-1}(x_t + x_{t+1}) = \tfrac{1}{2}\theta y_{t-1}^1 \tag{7b}$$

$$a_t = \tfrac{1}{2} E_{g,t}(x_t + x_{t+1}) = \tfrac{1}{2}(\rho \bar{y}_t + \theta y_t^1) \tag{7c}$$

where

$$\rho \equiv 2\sigma_x^2/(2\sigma_x^2 + \sigma_\varepsilon^2); \; \theta \equiv \sigma_x^2/(\sigma_x^2 + \sigma_\varepsilon^2); \; \bar{y}_t \equiv (y_t^0 + y_{t-1}^1)/2,$$

$$t = 1, \ldots, n.$$

---

11. Obviously the last period in which the public's welfare is affected by the policies of the incumbent government is period $n + 1$ through the term $(a_n - x_{n+1})^2$.

The unconditional expected value of the minimized objective function in each of the three cases is given by

$$EV_{t1}^m = E(x_t^2 + x_{t+1}^2) = 2\sigma_x^2 \tag{8a}$$

$$EV_{t2}^m = E\{[(\tfrac{\theta}{2})y_{t-1}^1 - x_t]^2 + [(\tfrac{\theta}{2})y_{t-1}^1 - x_{t+1}]^2\} = (2 - \tfrac{\theta}{2})\sigma_x^2 \tag{8b}$$

$$EV_{t3}^m = E\{[\tfrac{1}{2}(\rho\bar{y}_t + \theta y_t) - x_t]^2 + [\tfrac{1}{2}(\rho\bar{y}_t + \theta y_t) - x_{t+1}]^2\}$$

$$= \left[\frac{2-(\theta+\rho)}{2}\right]\sigma_x^2 \quad \text{and} \quad t = 1, \ldots, n \tag{8c}$$

Equations (8a) through (8c) show that the longer decisions are delayed, the smaller the loss of welfare, $EV_{t1}^m > EV_{t2}^m > EV_{t3}^m$. New information about states of the world is useful, so the best result is obtained when the social planner delays the decision regarding his policy in period t to the latest possible time, the start of period t.

Suppose that before taking office a social planner has to decide whether to commit policy instruments to particular values. Since the decision has to be made before the actual values of x_t and y_t^i, $i = 0, 1$ are known, he ranks alternatives (i) through (iii) using the information available at the time. This information includes only the deterministic and the stochastic structure of the economy, so the planner uses unconditional expected values to rank policies. Cases (i) and (ii) correspond to various degrees of precommitment of policy instruments, while case (iii) can be thought of either as a type of descretionary policy or as a contingent policy rule. Case (iii) requires the decision about the setting of the policy instrument for each period to be made in that period, but case (iii) results also if the planner commits himself to the contingent rule given by (7c).

Our standard for a benevolent planner is at hand. Social welfare is maximized either when the social planner has discretionary powers or when the planner follows a contingent rule that replicates his choice of instruments under discretion.[12] In either case, maximizing social welfare requires flexibility in the choice of settings or values for the policy instrument. Flexibility enables the planner to commit his instruments only after he has the maximum possible amount of information. In view of the foregoing discussion, the social planning problem in (3) can be rewritten

$$\min_{(a_1,\ldots a_n)} \sum_{t=1}^{n} E_{gt} V_t \tag{9}$$

The optimizing choice of instruments for this problem is given in (7c).

---

12. This result depends on the assumed "benevolence" of the planner. We show below that when the government has political aims, the social optimum can be achieved by a contingent rule but not by discretion.

III. A Politically Motivated Government with Partially Informed Voters

In models of political economy or public choice, policies are not chosen by benevolent planners. Policymakers maximize their own objective functions, which may differ from a well-defined social utility function. This section introduces a politically motivated government (or policymaker) and shows that such a government reduces social welfare below the standard set by the planner.

The government's main concern is to be reelected. The government knows that its prospects for reelection are directly related to its forecasting ability. The public believes that a government with better forecasting ability (lower σ_ε^2) is more likely to achieve higher social welfare if reelected.[13] Since the public does not know σ_ε^2, a_t, or x_t, its only source of information about the forecasting ability of the incumbent government is the level of welfare experienced during the period in office. We assume that the probability of reelection is given by

$$P[L(\underline{a}, \underline{x})], \quad P'(\cdot) \equiv \frac{\partial P}{\partial L} < 0 \tag{10}$$

where

$$\underline{a} \equiv (a_1, \ldots, a_n), \quad \underline{x} \equiv (x_1, \ldots, x_n)$$

and

$$L \equiv (a_0 - x_1)^2 + \sum_{t=1}^{n-1} [(a_t - x_t)^2 + (a_t - x_{t+1})^2] + (a_n - x_n)^2 \tag{11}$$

is the actual cumulative loss experienced by the public during the office period.

Let $u(R)$ and $u(NR)$ be the subjective levels of satisfaction experienced by government if reelected and if not reelected, respectively. Following Downs' (1957) view of the "political entrepreneur," we assume that $u(R) - u(NR) > 0$. The incumbent government chooses policies to win reelection by setting the vector of instruments a to maximize

$$\max_{a} E_g(P[(\underline{a}, \underline{x})]u(R) + \{1 - P[L(\underline{a}, \underline{x})]\}u(NR)) \tag{12}$$

Part 1 of the appendix shows that (using a linear approximation of $P(\cdot)$) this decision problem is equivalent to

$$\min_{a} E_g L(\underline{a}, \underline{x}) \tag{13}$$

---

13. We demonstrate later in this section that this belief is rational in the sense that the actual behavior of a politically motivated government facing such beliefs gives rise to a positive relationship, on average, between welfare during the period in office and the government's ability to forecast.

The government chooses the instruments, $\underline{a}$, to minimize the conditional expected value of the cumulative loss to the public during its term in office. As in Section II, government achieves better (lower) value of the objective in (13) on average if it delays the choice of a_t as long as possible. Using this consideration and (11), the problem in (13) can be rewritten

$$\min_{\underline{a}} \{(a_0 - x_1)^2 + \sum_{t=1}^{n-1} E_{gt}[(a_t - x_t)^2 + (a_t - x_{t+1})^2] + E_{gn}(a_n - x_n)^2\} \quad (14)$$

The first-order necessary conditions for an internal minimum yield

$$a_t = \tfrac{1}{2}(\rho \bar{y}_t + \theta y_t^1), \quad t = 1, \ldots, n-1 \quad (15a)$$

$$a_n = \rho \bar{y}_n \quad (15b)$$

where, as before, $\rho \equiv 2\sigma_x^2/(2\sigma_x^2 + \sigma_\varepsilon^2)$, $\theta \equiv \sigma_x^2/(\sigma_x^2 + \sigma_\varepsilon^2)$.

Comparison of (15) with the socially optimal setting of instruments in (7c) suggests that the instruments are set optimally in the first $n - 1$ periods but not in the last one. The reason for the divergence in the last period is that the public does not know the government's policy action, the choice of a_n. Consequently, the public does not have sufficient information to use (3) to evaluate policy. The best the public can do is to use information about performance. As a result, the government can improve its reelection prospects by setting a_n without regard for the effect of this choice of a_n on welfare in the period following the elections. The government sets $a_n = \rho \bar{y}_n$, lowering welfare. The loss of welfare is directly traceable to the existence of elections and an imperfectly informed public. The public knows that all governments try to appear better than they are. Using private forecasts or anticipating the government's action in period n cannot fully offset the advantage the government gets from having private information, however. The reason is that the government exploits private information for its own advantage. The public can try to evaluate government's performance by using, in addition to (11), the contribution of the choice of a_n to welfare in the first postelection period. This contribution is inversely related to the loss $(a_n - x_{n+1})^2$. The best forecast of this loss by the public prior to the election is $E_{pn}(a_n - x_{n+1})^2$, where the subscript pn denotes the information available to the public in period n. The public knows that a_n is set by the government according to equation (15b), so

$$E_{pn}(a_n - x_{n+1})^2 = E_{pn}[\tfrac{\rho}{2}(x_n + \varepsilon_n^0 + \varepsilon_{n-1}^1) - x_{n+1}]^2$$

An important feature of this expected value is that it depends on the joint distribution of x, ε^0, ε^1, and ρ but *not* on the actual choice of a_n, which is unknown to the public. Hence the government has no incentive to take welfare in period $n + 1$ into consideration when setting a_n. This leads to the policy choice in (15b). Despite the fact that the public is aware of government's tendency to act suboptimally prior to elections, a rational government acts in this way. A government that failed to sacrifice postelection welfare for preelection welfare would be judged less capable than it really is. Since the public believes that all governments tend to disregard postelection welfare, the public's evaluation of government's ability is based on this belief.[14]

Since the public cannot determine whether the losses it suffers in a given period are due to current or to past decisions, the incumbent government is penalized for the actions of its predecessor through the term $(a_0 - x_1)^2$ over which it has no control. On the other hand, it is not penalized for its contribution to losses in period $n + 1$, through the term $(a_n - x_{n+1})^2$, even though it affects this term by the choice of a_n. In addition, the incumbent government is penalized for bad luck (large deviations of ε and x from their respective means) and rewarded for good luck.

Substituting the instrument levels chosen in (15) into (11) and rearranging, we obtain the expected value of cumulative losses during the term of office arising from the government's attempt to maximize its reelection prospects,

$$EL^* = \left[1 + (1 + \rho_p) + 2(n - 1) - \frac{3 + (n - 1)(\rho + \theta)}{2}\right]\sigma_x^2$$

where ρ_p is the ρ ratio of the previous government. On average, cumulative welfare increases with the precision of the incumbent government's forecasts (the lower σ_ε^2 for a given σ_x^2). The equation shows that there is a positive relationship between σ_ε^2 and expected losses during a term of office. This establishes the rationality of the public's beliefs about this relationship. The public is correct to prefer a government with better forecasting ability; if elected, that government is more likely to achieve a better level of social welfare. Since the public does not know σ_ε^2, it uses the fact that actual welfare during the office period is positively related, on average, to the incumbent government's forecasting ability to evaluate the way the government will perform if reelected. As a result, the likelihood of reelection increases with the level of welfare generated during its term of office.

Government is aware of the relationship between reelection and welfare during its term, so it behaves in a way that sustains the relation. The public's beliefs are rational in the sense that the beliefs are not obviously controverted by observations.[15] Note that although the public is forward looking, it evaluates the government in terms of past performance, since this information is pertinent (given the informational limitations) for prediction of future performance. Reliance on past performance is consistent with a vast amount of literature

---

14. This is analogous to the theory of limit pricing under asymmetric information about the costs of an incumbent firm. Since potential entrants take the incumbent firm's price as a signal about its costs, the incumbent has an incentive to set its price below the profit maximizing price. The potential entrant who is aware of this incentive is not fooled on average. Nevertheless, the incumbent practices limit pricing because otherwise he would be judged to have costs that are higher than actual costs (Milgrom & Roberts, 1982; Roberts, 1985). We are indebted to Motty Perry for pointing out this analogy.

15. This rather general notion of rationality is due to Radner (1977).

suggesting that the reelection prospects of a government are better the better are economic conditions during its term.[16]

The public's welfare is reduced, however, by the government's concern about reelection. The average loss in social welfare resulting from the government's preelection activity and the public's imperfect information can be quantified. Substituting (7c) and (15b) alternately into (4) for $t = n$, and taking the expected value of the difference between the resulting expressions, we have the expected value of the loss.

$$E\{(\rho\bar{y}_n - x_n)^2 + (\rho\bar{y}_n - x_{n+1})^2 - [\tfrac{1}{2}(\rho\bar{y}_n + \theta y_n^1) - x_n]^2 - [\tfrac{1}{2}(\rho\bar{y}_n + \theta y_n^1) - x_{n+1}]^2\}$$
$$= (1 + \rho)\sigma_x^2 \qquad (16)$$

The loss of welfare is an increasing function of the uncertainty of the states of nature, σ_x^2; the higher this uncertainty, the more valuable from a social point of view the advance information that government has but does not use to achieve a social optimum. Conversely, the higher σ_ε^2 the less accurate is the advance information available to the government, so social welfare is not affected much by the neglect of this information.[17]

We have shown that maximizing behavior of a politically motivated government creates a social inefficiency in the presence of asymmetric information. The inefficiency is independent of the source of the asymmetry; the loss can arise either from slow dissemination of information by government or from slow learning or reception by the public, or from any combination of the two. If we introduced a challenger, the loss would not disappear as long as the government retained some informational advantage.

When the public uses realized welfare during incumbency as the criterion for the evaluation of future performance of the government, maximization of the likelihood of being reelected is incompatible with maximization of social welfare. The loss of social welfare is directly traceable to the existence of periodic democratic elections, so we have called it the cost of democracy.[18] Equation (16) suggests that this cost is higher the more uncertainty there is about future states of nature. Obviously the cost is also higher the shorter the election cycle.

---

16. Schneider and Frey (1984) survey this literature.

17. If voters (knowing that government tends to behave differently in the last period before the elections) give less weight to welfare in that period, the government will still have an incentive to set $a_n = \rho\bar{y}_n$ as long as this weight is nonzero. If voters totally discount the welfare of the last period, they give the government an incentive to set $a_{n-1} = \rho\bar{y}_{n-1}$, which creates an inefficiency in period n. More generally, any period that is excluded by the public when evaluating the government leads the latter to disregard losses in that period. The loss in welfare is not thereby diminished, since any period that is excluded from the public's evaluation causes an average increase in losses of $(1 + \rho)\sigma_x^2$. Since, from (16), this is equal to the average reduction in losses due to the exclusion of the last period from the public's evaluation, there is no gain in shifting the weights of the evaluation function across periods.

18. Our analysis does not show that democracy is inferior to other viable alternatives. Cost is measured from an ideal point, not from the point achieved under an alternative social arrangement. Further, the benefits of democratic government are neglected.

IV. A Politically Motivated Government with Fully Informed Voters

This short section locates the reason for the inefficiency that gives rise to the cost of democracy. We show that when everyone has the same information as the government, a politically motivated government sets all instruments at their socially optimal levels and the cost of democracy vanishes.

Suppose all members of the public know $a_1, \ldots, a_n, x_1, \ldots, x_n, y_1^0, \ldots, y_n^0$, and $y_1^1, \ldots, y_{n-1}^1$, in the last period before the elections. Using this information, they can estimate the forecasting ability of the incumbent government directly by noting that

$$y_t^0 - x_t = \varepsilon_t^0, \quad t = 1, \ldots, n \quad \text{and} \quad y_t^1 - x_{t+1} = \varepsilon_t^1, \quad t = 0, \ldots, n-1$$

and by using the series of observations on ε_t^0 and ε_t^1 to estimate the variance σ_ε^2 by means of a statistic like[19]

$$\sum_{t=1}^{n} \frac{(\varepsilon_t^0 + \varepsilon_{t-1}^1)}{2n}$$

The public can now separate the errors introduced by the government's forecasts, so prospects for reelection depend on a direct estimate of σ_ε^2 rather than on social welfare attained during the term of office.

The government no longer has an incentive to deviate from the socially optimal choice of instruments. When the public has complete information, a government that seeks reelection will choose to maximize social welfare. The incentive to maximize welfare remains only if *all* individuals are as informed as the government. When *some* individuals are not as knowledgeable, and therefore use actual welfare to evaluate government as in Section III, the incentive to deviate from social optimality reappears.

The departure from a social optimum, that we have called the cost of democracy, arises here from differences in information. Our analysis brings out the public good nature of information in a democratic society. The private benefit of information to an individual or a group of individuals may be lower than the private costs of acquiring the information even if the social benefits from more efficient behavior by government are larger than the costs.

V. Implementation of Constitutional Rules

Many economists advocate some form of precommitment, or rules, for governmental policies. Friedman (1960) and more recently Kydland and Prescott (1977) are examples. Yet governments usually oppose attempts to restrict their freedom to change policies. When there is uncertainty about future states of nature, governments that desire to maximize the probability of being reelected

---

19. This is a maximum likelihood estimate of the variance.

choose to retain as many open options as possible. If they do not behave in this way voters penalize them at election time.

In previous sections we showed that both politically motivated (Section III) and socially motivated (Section II) choices of policy instruments give rise to a demand for flexibility on the part of government. Flexibility is good for social welfare and is also good politics. There is a fundamental difference, however, between the type of institutional arrangements used to maintain flexibility in these two cases. Flexibility can achieve a social optimum either through discretion or by means of a contingent rule of the type specified in (7c). With an apolitical social planner who maximizes expected social welfare, it does not matter whether government is precommitted through an appropriate contingent rule or has discretionary powers. In either case social welfare is maximized. But with a politically motivated government, discretion leads to a socially suboptimal result while the contingent rule in (7c) achieves a social optimum. Hence a constitution that precommits the government to the contingent rule in (7c) is desirable. The familiar problem of enforcement arises, however.

The problem of implementing or enforcing a constitutional rule arises because the authority that enforces the rule must have the same information as the government to perform its function. In particular, it must know y_t^0 and y_t for each t. Such a complete transfer of information is usually not feasible, since y_t^i, $i = 0, 1$ represent governmental forecasts rather than realizations of objective variables. Even if a complete transfer of information can be made, neither the government nor the public may choose to make the transfer if y_t^i includes information used to set military strategy.[20] As a consequence, the implementation of a constitutional rule rests ultimately with the ability of the public and the incentives on government to release all available information.

If the government cares more about the current than the future election it will usually be tempted to abandon the constitutional rule in favor of the political behavior described by equation (15). To see this, consider first the extreme case in which government cares only about the next election. As in Section III, the public evaluates the forecasting ability of government on the basis of welfare experienced during its period in office, since this measure of welfare is positively correlated with forecasting ability. By following the behavior in equation (15), the government increases the quality of its forecasting ability as perceived by the public and enhances the likelihood of reelection. This behavior is optimal for reelection prospects whether the public believes that the government adheres to the constitutional rule or is fully or partially aware of the fact that government acts in a discretionary manner. Whatever the public's beliefs about the policy regime, government is perceived as having better forecasting ability when it sets $a_n = \rho \bar{y}_n$ than when it sets a_n at the socially optimal level implied by equation (7c). Thus $a_n = \rho \bar{y}_n$ is a dominant strategy from the point of view of government.

20. Defense differs from economic policy because the social cost of distributing information can be larger than the social gain from announcing and following a precommitted plan. The social cost arises from the revelation of forecasts of strategic variables or enemy responses. Such forecasts reveal strategies and interpretations to enemies or potential enemies.

This tendency is attenuated to some extent if the government also cares about future elections. In this case the government knows that, after the election, the public will learn about the loss of welfare resulting from the government's choice of a_n. In the subsequent election, the public's loss of welfare will be costly to the government. Nevertheless, to the extent that the government cares about the present election sufficiently more than about the subsequent election, the temptation to abandon the constitutional rule remains. Even if the public and the government have an identical degree of time preference this is likely to be the case. The reason is that discretionary behavior reduces the public's welfare immediately after the election, but the penalty to the reelection prospects of government is delayed until the next election. More generally, the bias toward discretion remains whenever the cost of democracy is internalized by government later than by the public.[21]

Thus even if a constitution is enacted, government is tempted to violate it whenever the private costs to government occur at a later date (or more generally are lower) than the costs of discretion to the public. The temptation to act in a discretionary manner is directly traceable to the existence of asymmetric information. If the public had the same information as government, it could estimate the forecasting ability of the latter directly along the lines of Section IV, without having to rely on actual welfare during the period in office. If this occurred, there would be no incentive for government to deviate from the socially optimal behavior given by (7c). When only part of the public is informed, the bias toward discretion reappears. The reason is that the uninformed judge the forecasting ability of government by the level of welfare generated during its time in office. With periodic elections, information is a public good in the sense that the private net benefits of becoming informed are, for some people, lower than the social benefits.

VI. An Example

This section provides an application that maps into the quadratic loss function discussed in previous sections. In this example the random state variable x is the productivity of labor in the production of public good, and the government's policy instrument is the amount of labor drafted to produce the good. The distribution of x and the structure of information remain as in Section II.

The amount of the public good produced in any period depends on the amount of labor input in the current and the previous period. Labor productivity in the production of the public good is stochastic and not known with certainty in advance. Government obtains labor by directly drafting labor to produce the public good.[22] An example is a military draft to provide defense.

---

21. In practice, another reason for the bias is that an incumbent may not be able to pass his good reputation onto his successor as the party's candidate.

22. The assumption about a labor draft is made to abstract from the distortions created when the public good is financed by taxation. The assumption has greatest descriptive realism when the public good is defense.

Since the amount of labor currently drafted effects the level of defense provided in both the current and the next period, governments with superior ability to forecast make more efficient decisions. The reason is that a government with better forecasts uses more labor when productivity in this sector is relatively high and less when productivity is relatively low. In countries like Israel, where defense is based on civilian reservists, a government with good forecasting ability drafts reservists only when danger is imminent. In such periods, productivity in the public good sector is relatively high.

To formalize these notions, we assume that the production function of the public good is

$$g_t = f(Nd_{t-1}, Nd_t, x_t) \tag{17}$$

where g_t, d_t, x_t, and N are, respectively, the amount of the public good provided in period t, the amount of labor drafted per individual in period t, the productivity shock in period t, and the constant population. Utility of the jth individual is given by

$$u^j(c_{tj}, l_{tj}, g_t) \tag{18}$$

where c_{tj} and l_{tj} are respectively the jth individual's consumption of the private good and his leisure in period t. Each individual is endowed with one unit of time per period. Part of his time is drafted by government, and the rest is distributed between leisure and private production. Each unit of labor in private production produces one unit of the private consumption good. Hence the time constraint of a representative individual can be written

$$l_{tj} = 1 - c_{tj} - d_{tj}, \qquad j = 1, \ldots, N \tag{19}$$

The total level of drafting is determined by the government, whereas its distribution is determined by the specific institutions in charge of the production of the public good. More precisely, the level of drafting of individual j's time in period t is

$$d_{tj} = d_t(1 + V_{tj}) \tag{20}$$

where

$$V_{tj} \geq -1 \quad \text{for all } j \quad \text{and} \quad \sum_{j=1}^{N} V_{tj} = 0 \tag{21}$$

It follows from (20) and (21) that total labor drafted in period t is

$$\sum_{j=1}^{N} d_{tj} = Nd_t \tag{22}$$

The total number of draftees is determined by the government's choice of d_t. The distribution of the drafting effort is determined by technical considerations outside the range of governmental jurisdiction so that the V_{it} are not governmental policy variables. When the public good is defense, this means that the total number drafted is determined by the central government. Given this level, the allocation of the burden is determined by other institutions (the army or draft boards). Individual j knows his own condition, d_{tj}, but cannot use this information to infer the average level of d_t, since V_{tj} is not known to him. Hence as in Section III, the settings of the government policy instruments are private information.

Given the government's forecast and its current and past decisions, the representative individual maximizes (18) subject to (17) and (19) by solving the following private optimization problem,

$$\max u^j[c_{tj}, 1 - c_{tj} - d_{tj}, f(Nd_{t-1}, Nd_t, x_t)] \tag{23}$$

The result of the optimization is an indirect utility function that depends on d_{tj}, d_t, d_{t-1}, and x_t and that is given by[23]

$$I^j(d_{tj}, d_{t-1}, d_t, x_t) \tag{24}$$

As in Section III, the public evaluates the government on its performance during its term of office. The government's problem is to choose values of d that maximize

$$\max_{\{d_t, t=1,\ldots,n\}} \sum_{t=1}^{n} E_{gt} I(d_{t-1}, d_t, x_t) \tag{25}$$

where

$$I(d_{t-1}, d_t, x_t) \equiv \sum_{j=1}^{N} W_j I^j[d_t(1 + V_{tj}), d_{t-1}, d_t, x_t] \tag{26}$$

and

$$W_j \geq 0 \tag{27}$$

$$\sum_{j=1}^{N} W_j = 1 \tag{28}$$

The W_j's are nonnegative weights that reflect the electoral power of individuals of type j.[24] The maximization problem in (25) maps into the problem solved by government in equation (14) for a particular specialization of the indirect utility function $I(\cdot)$. Specifically, for

$$I(d_{t-1}, d_t, x_t) = A - (d_{t-1} - x_t)^2 - (d_t - x_t)^2 \tag{29}$$

---

23. Since N is constant, it is subsumed in the function.
24. Since the government takes the distribution parameters V_{tj} as given when choosing d, they are subsumed into the function $I(\cdot)$.

and

$$d_t \equiv a_t \quad \text{for all } t \tag{30}$$

government's problem in (25) reduces to the minimization problem in (14).

VII. A Politically Motivated Government Facing Voters with Diverse Objectives

To this point we have assumed that all individuals have an identical loss function, and we have used the loss function of the representative individual as a measure of social loss. This section relaxes that restriction by permitting individual loss functions to differ. Governmental decisions now affect both total welfare and its distribution across individuals. A government seeking reelection sets its instruments to determine the distribution of welfare across individuals so as to maximize the probability of reelection.

Individuals differ in their preference for activist policy. "Activists" prefer large responses by government to deviations of the state variable, x, from its mean; they favor large changes in $\underline{a}$ and, ceteris paribus, they vote to reelect governments that are "responsive." "Nonactivists" prefer small to large changes whenever x deviates from its mean.

Let β_W be the weight assigned by individuals with different preferences for activist policy. The loss function of an individual of type w in period t is

$$L_{tw} \equiv (a_{t-1} - \beta_w x_t)^2 + (a_t - \beta_w x_t)^2 \tag{31}$$

As in Section III, individuals do not directly observe a_t, x_t, and the degree of foresight of the incumbent government as summarized by σ_ε^2. Each person knows the level of welfare he experiences during the term of office and also has self-fulfilling beliefs about the relationship between σ_ε^2 and the expected value of his own welfare. Each person uses welfare experienced during the incumbent's term to form a rational forecast of his own welfare if the incumbent government is reelected. The likelihood that an individual or group votes for reelection increases with his level of welfare. As before, the probability of reelection is given by equation (10), but equation (11) is replaced by (32).

$$L = \sum_{w=1}^{W} \psi_w \sum_{t=1}^{n} L_{tw} \tag{32}$$

where W is the number of different types of individuals and ψ_w is a coefficient that measures the marginal effect on the probability of reelection of the cumulative welfare experienced by group w during incumbency. Let λ_w be the weight of group w in the population. Obviously

$$\sum_{w=1}^{W} \lambda_w = 1 \tag{33}$$

and ψ_w is an increasing function of λ_w. For simplicity and without loss of generality, we set

$$\psi_w = \lambda_w \quad \text{for all } w \tag{34}$$

Substituting (31) and (34) into (32) and using the same considerations as those that led to equation (14), it follows that maximization of the probability of reelection[25] is equivalent to

$$\min_a \left\{ \sum_{w=1}^{W} \lambda_w (a_0 - \beta_w x_1)^2 + \sum_{t=1}^{n-1} E_{gt} \sum_{w=1}^{W} \lambda_w [(a_t - \beta_w x_t)^2 + (a_t - \beta_w x_{t+1})^2] + E_{gn} \sum_{w=1}^{W} \lambda_w (a_n - \beta_w x_n)^2 \right\} \tag{35}$$

The first-order necessary conditions for this problem yield

$$a_t = \frac{\bar{\beta}}{2}(\rho \bar{y}_t + \theta y_t^1), \quad t = 1, \ldots, n-1 \tag{36a}$$

$$a_n = \bar{\beta} \rho \bar{y}_n \tag{36b}$$

where

$$\bar{\beta} \equiv \sum_{w=1}^{W} \lambda_w \beta_w \tag{36c}$$

$\bar{\beta}$ is the (weighted) mean value of the activism parameter for individuals in the population.

The government sets its instruments as if the population was composed of identical individuals, all of whom share the preference $\bar{\beta}$ for activism.[26] As before, the government does not choose the socially optimal value for the policy instrument in the last period before the election; it chooses $\bar{\beta} \rho \bar{y}_n$. To show that this value is not socially optimal, it is necessary to redefine social optimality in terms of Pareto efficiency since, with differences in tastes, many combinations of welfare are consistent with social efficiency.[27] The Pareto-efficient frontier is derived by solving the following minimization problem:

$$\min_a \sum_{w=1}^{W} \lambda'_w \sum_{t=1}^{n} E_{gt}[(a_t - \beta_w x_t)^2 + (a_t - \beta_w x_{t+1})^2] \tag{37}$$

---

25. This is the same as maximization of government's objective function in (12).

26. When the distribution of β is symmetric, the mean $\bar{\beta}$ is equal to the median and government behaves as if the median voter was an imperfectly informed dictator. This result is similar to that obtained by Meltzer and Richard (1981) in the context of income redistribution with perfectly informed voters.

27. In this contest, Pareto efficiency means that it is not possible to improve the welfare of any group of individuals while holding the level of welfare of all the other groups constant.

where the λ'_w are arbitrary positive weights that sum to one. The first-order conditions for the problem in equation (37) are

$$a_t = \frac{\bar{\beta}'}{2}(\rho \bar{y}_t + \theta y_t^1) \tag{38}$$

for all t and, in particular, for $t = n$. Here $\bar{\beta}'$ is the weighted mean value of the β_w using the weights λ'_w.

Comparison of equations (36b) and (38) shows that in the last period before the elections the policymaker chooses a value of a_n that drives the economy inside the Pareto-efficient frontier. This result is true in general and in particular when $\bar{\beta}' = \bar{\beta}$. As in Section III, the government takes advantage of its superior information at the time of the election to increase the probability of reelection despite the cost to society. The cost, or inefficiency, takes the form of excessive interest in the welfare of the mean voter in period n and complete disregard of the effect on his welfare in the first period after the election.

To evaluate the expected welfare of voter s when government seeks reelection, we compute the expected value of

$$L_s \equiv \sum_{t=1}^{n} L_{ts} \tag{39}$$

for a government that seeks to maximize the probability of reelection. Substituting (36a) and (36b) into (31), substituting the resulting expressions into (39), taking expected values and rearranging, we obtain

$$EL_s = \left\{ k_s + \bar{\beta}\left(\frac{\bar{\beta}}{2} - \beta_s\right)[(n+1)\rho + (n-1)\theta] \right\} \sigma_x^2 \tag{40}$$

where

$$k_s \equiv 2\left[n\beta_s^2 + \bar{\beta}\left(\frac{\bar{\beta}}{2} - \beta_s\right)\rho_p\right]$$

Equation (40) divides the population into three groups based on the relative size of β_s. People with $\beta_s = \bar{\beta}/2$ are unaffected by the government's ability to forecast as measured by σ_ε^2. Those with $\beta_s > \bar{\beta}/2$ gain from improved forecasts (lower σ_ε^2). The expected value of their loss, EL_s, declines as σ_ε^2 increases. This group includes people with relatively high values of β. Such people favor activist policies, so they prefer government action to inaction. Improved forecasting ability increases policy activism, so the gain from improved forecasting is reinforced by the gain from increased activism. For the remaining group, with $\beta_s < \bar{\beta}/2$, the two effects work in opposite directions. Greater precision decreases average losses of the mean voter and may even increase the expected welfare of voters that are not far from the mean. On the other hand, government becomes more activist on average. Increased activism reduces the welfare of all voters with value of β below the mean. For those sufficiently close to the mean, the first, positive, effect is dominant. For voters with a sufficiently strong aversion toward activism, the second effect dominates, and their expected welfare is lower when the incumbent government is more precise in its forecasts.

Each group uses the actual level of welfare experienced during the government's term of office as an indicator of its expected welfare during a new term. This practice is rational. Actual welfare is an indicator of σ_ε^2, and σ_ε^2 determines expected welfare if the government is reelected.

The effect of σ_ε^2 on groups with relatively low and high β_s differs, however. Strong activists vote for reelection of a government with good forecasting ability, while strong antiactivists vote against. The latter are a minority. The combined effect, over all groups, is usually positive; governments with better forecasting performance have a better chance to be reelected. This is shown by calculating the conditional expected value of (32) when a government seeks reelection. Using equations (33), (36c), (39), and (40) in (32) and rearranging

$$EL = \left\{ \sum_{w=1}^{W} \lambda_w k_w - [(n+1)\rho + (n-1)\theta] \frac{\bar{\beta}}{2} \right\} \sigma_x^2$$

It follows that (for $n > 1$) governments with better forecasting ability are more likely to be reelected.

VII. Concluding Remarks

In a democracy with periodic elections and an imperfectly informed public, discretionary policy does not lead to a socially optimal choice of policy instruments. We call the resulting loss the cost of democracy. The cost is a gross cost; we make no attempt to compare democracy to other systems, and we neglect all the benefits of democratic government.

The particular cost of democracy that we have identified arises because the public has less information than the government. An optimal state contingent policy rule that precommits government through a constitution, or a similar device, eliminates the cost of democracy by removing the government's opportunity to exploit its information advantage. There are problems of implementation, however, The optimal constitution implied by our model makes governmental actions contingent on the advance information available only to government. Enforcement of a constitutional rule requires full and current knowledge of governmental forecasts. The implementation of the constitution depends, therefore, on the government. Government is tempted not to abide by the constitution when it can improve reelection prospects.

The cost of democracy disappears, without a constitution, when *all* the public are fully informed. Full information by a relatively small group is not sufficient to wipe out the cost, however. The cost reemerges whenever a substantial fraction of voters is uninformed.

We have used a particularly simple framework that abstracts from discounting of future welfare and restricts the length of the lag between policy actions and their effects on welfare. Also, we have used a particular hypothesis to show that monitoring is costly and to give the government an advantage in information. Obviously, there are other sources of informational advantage and other reasons for high costs of monitoring than those we have used.

Our model implies that any government with private information that maximizes the probability of reelection will choose not to maximize social welfare. The public expects the government to increase its welfare before an election, at the expense of a greater loss of future welfare, and it judges the government's competence by its performance in advance of the election. A failure of the government to act in its own interest before the election gives an incorrect inference to the public about the government's competence—measured here by its ability to forecast the performance of the economy.

An example, discussed in Section VI, illustrates the qualitative properties of the model in situations where flexibility can increase social welfare. When there are a number of uninformed voters and policy is discretionary, the government's behavior prior to an election is suboptimal. The government uses discretion to increase the probability of reelection at the cost of diminished welfare after the election.

The principles demonstrated have wider application than the specifics of the example used to illustrate the argument or the quadratic loss function and other hypotheses used to develop the model. We believe the principles apply to any area in which current governmental decisions affect welfare beyond the current period and in which there is asymmetric information.

Even with full transmission of information, a constitution of the type considered here requires unanimity with regard to social objectives. When, as in Section VII, individuals differ in their views concerning the optimal degree of activism, it may be difficult to agree on the form of the constitution. Making the constitution contingent on the relative power of various groups reduces certainty about future policy, a main advantage of the constitutional rule. Divergence of views within society, differences in information, and the electoral advantage to the government from witholding information appear to be sufficiently common to explain why most government policy remains discretionary.

Appendix

Derivation of Equation (13)

Since $u(R) - u(NR) > 0$, the maximization problem in (12) is equivalent to

$$\max_{a} E_g P[L(\underline{a}, \underline{x})] \tag{A1}$$

Approximating $P[L(\underline{a}, \underline{x})]$ linearly around the point $a = x = 0$

$$P[L(\underline{a}, \underline{x})] = P[L(0, 0)] + P'[L(0, 0)][L(\underline{a}, \underline{x}) - L(0, 0)] \quad \text{(A2)}$$

Since from (11), $L(0, 0) = 0$, (A2) can be written

$$P[L(\underline{a}, \underline{x})] = K_0 - KL(\underline{a}, \underline{x}) \quad \text{(A3)}$$

where $K_0 \equiv P[L(0, 0)]$ and $K \equiv |P'[L(0, 0)]|$. Substituting (A3) into (A1)

$$\max_{\underline{a}} E_g P[L(\underline{a}, \underline{x})] = K_0 - K \min_{\underline{a}} E_g L(\underline{a}, \underline{x})$$

from which it follows that the maximization problem in (A1) is equivalent to the minimization problem in (13). This establishes the equivalence between the latter problem and (12).

References

Bernanke, B. S. (1983). "Irreversibility, Uncertainty and Cyclical Investment." *Quarterly Journal of Economics*, 97 (Feb.), pp. 85–106.

Canzoneri, M. (1985). "Monetary Policy Games and the Role of Private Information." *American Economic Review*, 76 (Dec.), pp. 1056–70.

Cukierman, A. (1980). "The Effects of Uncertainty on Investment under Risk Neutrality with Endogenous Information." *Journal of Political Economy*. 88 (June), pp. 462–75.

Downs, A. (1957). *An Economic Theory of Democracy*, New York: Harper.

Fischer, S. and Huizinga, J. (1982). "Inflation, Unemployment, and Public Opinion Polls." *Journal of Money Credit and Banking*, 14 (Feb.), pp. 1–19.

Friedman, M. (1960). *A Program for Monetary Stability*. New York: Fordham University Press.

Frey, B. S. and Schneider, F. (1978). "An Empirical Study of Politico-Economic Interactions in the U.S." *Review of Economics and Statistics*, 60 (May), pp. 174–83.

Kydland, F. E. and Prescott, E. C. (1977). "Rules Rather than Discretion: The Inconsistency of Optimal Plans." *Journal of Political Economy*, 85 (June), pp. 473–92.

Lucas, R. E. Jr. (1973). "Some International Evidence on Output Inflation Tradeoffs." *American Economic Review*, 63 (June), pp. 326–35.

Lucas, R. E. Jr. (1975). "An Equilibrium Model of the Business Cycle." *Journal of Political Economy*, 83 (Dec.), pp. 1113–44.

MacRae, D. (1977). "A Political Model of the Business Cycle." *Journal of Political Economy*, 85, pp. 239–63.

McCallum, B. T. (1978). "The Political Business Cycle: An Empirical Test." *Southern Economic Journal*, 44, pp. 504–15.

Meltzer, A. H. and Richard, S. F. (1981). "A Rational Theory of the Size of Government." *Journal of Political Economy*, 89, pp. 914–27.

Meltzer, A. H. and Vellrath, M. (1975). "The Effects of Economic Policies and Votes for the Presidency: Some Evidence from Recent Elections." *Journal of Law and Economics*, 18, pp. 781–98.

Milgrom, P. and Roberts, J. (1982). "Limit Pricing and Entry Under Incomplete Information: An Equilibrium Analysis." *Econometrica*, 50 (Mar.), pp. 443–59.

Nordhaus, W. (1975). "The Political Business Cycle." *Review of Economic Studies*, 42, pp. 169–90.

Radner, R. (1977). "Existence of Equilibrium with Imperfect Market Models." IMSSS, Stanford University, Stanford, Calif., July.

Roberts, J. (1985). "Battles for Market Share, Incomplete Information, Aggressive Strategic Pricing, and Competitive Dynamics." Paper presented at Symposium on Sequential Game Theoretic Models of Industrial Competition and Bargaining, Congress of the Econometric Society, August.

Schneider, F. and Frey, B. S. (1984). "Public Attitudes towards Inflation and Unemployment and their Influence on Government Behavior." Unpublished Manuscript. Pittsburgh: Carnegie-Mellon University, Graduate School of Industrial Administration.

8

A Theory of Ambiguity, Credibility, and Inflation under Discretion and Asymmetric Information

ALEX CUKIERMAN AND ALLAN H. MELTZER

I. Introduction

Central banks usually do not follow well-specified policy rules. Instead they move policy variables to reflect their changing emphasis on objectives such as high employment and low inflation. As emphasized by Weintraub (1978), Woolley (1984), and Hetzel (1984), the Federal Reserve is influenced by Congress, interest groups, and the electorate. Arthur Burns (1979) appears to have shared this view. He claimed that the Federal Reserve can work to achieve price stability only if the policy does not create too large an adverse effect on production and employment and does not irritate the Congress, the body to which the Federal Reserve is formally responsible. In Burns' words, the role of the Fed is to continue "probing the limits of its freedom to undernourish ... inflation" (Burns, 1979, p. 16). Evaluation of those limits involves a judgmental process, which uses as inputs both economic and political factors that shift in an unpredictable manner over time and about which the public has less timely information than the monetary authority.

This chapter explores the implications of this informational advantage for inflation and the credibility of policymakers when the latter maximize their own objective functions. A central objective of the chapter is to establish conditions under which ambiguity and imperfect credibility are preferable from the point of

We would like to thank without implicating an anonymous referee for perceptive reading of an earlier draft. Ernst Baltensperger, Ben Bernanke, Karl Brunner, Allen Drazen, Benjamin Eden, George Evans, Stephen Goldfeld, Marvin Goodfriend, Ed Green, Peter Hartley, Robert Hetzel, Bennett McCallum, Kenneth Rogoff, Thomas Sargent, John Taylor. Joseph Zeira and Itzhak Zilcha made helpful comments on earlier drafts of this chapter. Dave Santucci and Uriel Wittenberg provided efficient computational support. A previous version of this chapter was presented at the summer 1984 meeting of the Econometric Society, Stanford, California, at the Tel-Aviv Conference on Economic Policy in Theory and Practice and at the Konstanz Seminar.

view of policymakers to explicit formulation of objectives and perfect credibility.[1] The strong penchant of the Federal Reserve for secrecy is revealed in the legal record of a case in which the Federal Open Market Committee (FOMC) was sued under the Freedom of Information Act of 1966 to make public immediately after each FOMC meeting the policy directives and minutes for that meeting (Goodfriend, 1986). The Federal Reserve resisted and argued the case for secrecy on a number of different grounds. The present chapter provides a theoretical explanation for the Federal Reserve's desire for secrecy and ambiguity. We believe the analysis is applicable to other central banks.

The chapter also analyzes the quality of monetary control. Imperfect control of money increases opportunities for ambiguity, because the public is less able to detect shifts in policy when they occur. A policymaker who desires less than perfect credibility can reduce the quality of monetary control by avoiding efficient control procedures. We relate the quality of monetary control and the credibility of the policymaker and show that policymakers do not necessarily choose the most efficient control procedure that is technologically available. A certain degree of ambiguity enables policymakers to stimulate economic activity when they care most about such stimulation. Ambiguity enables policymakers to create monetary surprises when stimulation via surprises is most advantageous to the policymaker. As a result the optimal level of political ambiguity in the conduct of monetary policy may be larger than the minimum that is technologically attainable.

The chapter also investigates the relationship between the structure of policymakers' objectives and their credibility. Credibility is characterized in terms of the speed with which the public is convinced that new policies have been adopted when this is the case.

Recent literature on credibility and reputation has focused on finding conditions under which a shift to a policy rule as opposed to discretion is sustainable and therefore credible (Barro & Gordon, 1983a; Taylor 1982). The motivation of this literature is normative. The present chapter focuses on positive rather than normative issues. The maintained assumption is that the policymaker is free to follow discretionary policies. This seems to be roughly consistent with the actual state of affairs in the United States and other countries. The Federal Reserve and other central banks are not bound by any particular monetary rule.

Although the public forms expectations rationally, it cannot, in the presence of noisy monetary control, distinguish perfectly between persistent shifts in the setting of policy instruments and transitory control deviations. A consequence of the inability to separate these persistent and transitory shifts is that low credibility can persist for a time, particularly when objectives change substantially. The model also permits us to parametrize credibility as a continuous variable whose time path and laws of motion depend on the underlying

---

1. This view of choice of policies by the monetary authority is consistent with the observation by students of the Fed like Brunner and Meltzer (1964), Lombra and Moran (1980), and Kane (1982) that monetary policy is not formulated within a precise analytical framework.

characteristics of the policymaker's objective function and on the quality of monetary control.

Casual empiricism suggests that individuals use resources to monitor Federal Reserve actions and use current reports on the money supply as an indicator of future policies (Bull, 1982). This behavior is consistent with the thesis here—that expectations of future monetary growth react to actual monetary trends. A consequence of the public's behavior is that the policymaker takes into consideration the effect of current policy actions on future expectations even under discretion. However, since the policymaker has positive time preference, he partly discounts the effects of current policies on future expectations.

The model is consistent with observations showing that planned rates of monetary growth vary substantially both over time[2] and across countries. Other implications of the theory are:

1. Policy has an inflationary bias in the presence of control errors. Looser control of money raises the average rate of monetary growth above the rate required for price stability and decreases the credibility of shifts to new policies.

2. Monetary variability is not minimized. The variance of monetary growth is larger the higher the degree of time preference of the policymaker and the looser his control of money.

3. For a given quality of monetary control, the level of monetary uncertainty inflicted on the public increases with the time preference of the policymaker.

4. When policymakers are free to determine the accuracy of monetary control, they do not always choose the most effective control available. A deliberate choice of ambiguous control procedures is more likely the larger is the uncertain component of the policymaker's objectives in comparison with the certain component. We refer to countries with relatively large uncertainty in objectives as politically unstable.

5. The larger the political instability, the larger both the mean and the variance of monetary growth. This result provides a theoretical underpinning for the well-documented cross-country positive correlation between the level and the variability of inflation (Logue & Willet, 1976; Okun, 1971).

6. A policymaker is more likely to make a deliberate choice of ambiguous control procedures the higher his rate of time preference.

Taylor (1982) and Barro and Gordon (1983a) have argued that in the absence of constitutional enforcement, policy rules, as opposed to discretion, will not be credible because they are not dynamically consistent. There is no problem of dynamic inconsistency[3] here, and the absence of a binding

---

2. Large, frequency changes in planned monetary growth may be required by actual policy procedures. See Brunner and Meltzer (1964).

3. Section VI discusses this issue. We define dynamic inconsistency as the circumstance in which a dynamic optimizer, based on his objective function, (a) sets both policy for period t (today) and also sets contingent policies for future periods, but (b) actual policy in period $t + k$ differs from planned policy in $t + k$ set at time t. This definition is taken from the important paper by Kydland and Prescott (1977, pp 475–76). Kydland and Prescott also introduce an example (pp. 477–80) using a

constitutional rule is not the reason for imperfect credibility. Policy is discretionary, and credibility is imperfect because the policymaker has imperfect control of money, changing objectives, and an incentive to maintain some degree of ambiguity.

Section II presents the model and the policymaker's objective function. Section III shows the process by which the public forms its perceptions about the setting of the government's policy instruments and discusses the definition and the determinant of credibility under discretion. Section IV derives the optimal strategy of the policymaker and demonstrates the rationality of expectations. The implications for the distribution of money growth appear in Section V. In Section VI, we compare our model and results to other literature. Section VII derives the politically optimal level of accuracy in monetary control when this level can be altered by appropriate choice of institutions. The main result of this section is that the policymaker does not necessarily choose maximum accuracy. This is followed by concluding remarks.

II. The Model—General Structure

This section develops a framework for analyzing the choice of monetary growth by a policymaker whose objectives are described by a multiperiod, state-dependent objective function. Each period the policymaker chooses the rate of money growth so as to maximize the value of this function. When making choices the policymaker uses information that the public does not have: He knows the state in which his own objective function is.

The public knows the structure of government's decision-making process, but it does not know with certainty the state in which the government's objective function lies. The public forms a rational expectation about the current rate of money growth using information about past money growth.

The policymaker's objective function summarizes the government's preferences as a political entity. The government is concerned about public support and desires to stay in power.[4] We hypothesize that the government perceives that its chance of staying in power depends on the level of economic activity and on the rate of inflation. This view is consistent with public opinion polls showing that public support for the government rises with economic stimulation and decreases with the rate of inflation. (Fischer & Huizinga, 1982; Frey & Schneider, 1978; Goodhart & Bhansali, 1970).

---

model that is not explicitly dynamic and in which consistent and optimal policies differ. Some readers of our chapter have interpreted "dynamic inconsistency" in terms of this example. Our chapter corresponds conceptually to Kydland and Prescott's consistent solution but in an explicitly stochastic and dynamic framework. Credibility in our framework refers to changes in the policymaker's objectives within a discretionary and dynamically consistent framework.

4. Note that the objective function is *not* a social welfare function of the type used by economists (or planners) to reach normative judgments about optimal policy. We use the terms *government* and *policymaker* interchangeably, and do not relate the objective function to the behavior of voters or the policymaker's perception of the shifting weights voters place on inflation and unemployment.

The objective function is state dependent. The relative importance assigned to inflation and stimulation shifts in unpredictable ways as individuals within the decision-making body of government change their positions, alliances, and views. The changing weights may also reflect annual changes in the composition of a committee like the Federal Open Market Committee.[5] We assume, as suggested by Burns (1979), that the policymaker has more information about the timing of shifts than does the public.

The model is rational in the sense that the actual behavior that emerges is the same as the behavior on which the public relies to form expectations about future money growth. The government knows how the public forecasts monetary growth and inflation, so it can calculate, up to a random shock, the effect of a given choice of monetary growth on surprise creation. The government chooses the rate of money growth by comparing the benefits (to the government) from surprise creation against the costs of higher inflation.

Each period the policymaker plans to achieve a particular rate of monetary growth, m_i^p. Actual money growth, m_i, may differ from the planned rate because control is imperfect. More specifically

$$m_i = m_i^p + \psi_i \tag{1}$$

where ψ_i is period i's realization of a stochastic serially uncorrelated normal variate with zero mean and variance σ_ψ^2. This variance reflects the extent to which the operating procedures and the institutional environment prevent perfect control of money growth. At this stage σ_ψ^2 is taken as a technological parameter.

The policymaker's decision-making strategy is

$$\max_{\{m_i^p, i=0,1,\ldots\}} E_{G0} \sum_{i=0}^{\infty} \beta^i \left[e_i x_i - \frac{(m_i^p)^2}{2} \right] \tag{2}$$

$$e_i \equiv m_i - E[m_i | I_i] \tag{3}$$

$$x_i = A + p_i \qquad A > 0 \tag{4}$$

$$p_i = \rho p_{i-1} + v_i, \qquad 0 < \rho < 1 \tag{5}$$

where v is a serially uncorrelated normal variate with zero mean and variance σ_v^2 and is distributed independently of the control error ψ. Here e_i is the unanticipated rate of money growth in period i, I_i is the information available to the public at the beginning of period i, and $E[m_i | I_i]$ is the public's forecast of m_i given the information set I_i. The information set I_i includes all past values of m_i up to and including period $i - 1$. β is the government's subjective discount

---

5. The fact that committee membership does not change completely at any one time would then be consistent with the persistence of objectives introduced below. Changes in the weights that the Federal Open Market Committee has assigned to inflation or real income and to achieving monetary targets can be inferred from discussions in consecutive issues of the *Federal Reserve Bulletin*.

factor, and E_{G0} is a conditional expected-value operator that is conditioned on the information available to government in peiod 0, including a direct observation on x_0. Ceteris paribus the policymaker prefers lower to higher inflation.

x_i is a random shift parameter that determines the shifts in the policymaker's objectives between economic stimulation achieved through surprise creation and inflation. The higher the x_i, the more willing the policymaker to trade higher inflation for more stimulation. Equations (4) and (5) specify the stochastic behavior of the shift parameter x_i and indicate that governmental objectives exhibit a certain degree of persistence that depends on the size of A and ρ.

The probability that x_i will have a positive realization is larger than the probability that it will have a negative realization.[6] This reflects the hypothesis that unanticipated monetary growth stimulates employment and output[7] and that, ceteris paribus, the policymaker is more likely to prefer more to less stimulation. Since changes in the level of economic activity and in the rate of inflation have redistributional consequences,[8] shifts in x_i may originate from changes in the relative importance that the policymaker attaches to the welfare of various groups.

The public does not observe x_i directly but can draw inferences about the policymaker's objectives from observations of past money growth. However, inferences are not perfect because actual money growth reflects both the (persistent) plans of the policymaker and transitory control deviations.

Our formulation of the objective function reflects the basic uncertainty confronting the public. Society (including the policymaker) frequently experiences unanticipated events. Policymakers respond to these events by changing the relative weights that they place on the two objectives—stimulation and inflation. The policymaker does not reveal his current objectives or the trade-offs confronting him, so the public has less information than the policymaker about the relative weight currently assigned to surprise creation.[9]

The policymaker knows the process by which the public forms its perception, $E[m_i|I_i]$, of the current rate of monetary expansion. This process must

---

6. The distributional assumptions on v imply that $x \sim N(A, \sigma_p^2)$, where $\sigma_p^2 \equiv \sigma_v^2\sqrt{1-\rho^2}$. The statement in the text follows from the positive value of A and the fact that $\rho < 1$. As A increases in relation to σ_p^2, the probability of negative $x_i - s$ becomes smaller. For example, when $A = \sigma_p$ this probability is 0.16, and when $A = 2\sigma_p$ it is less than 0.03.

7. This effect can operate through any of the following: (a) via the mechanism described in Lucas (1973), (b) via nominal contracts of the type analyzed by Fischer (1977) and Taylor (1980a), (c) through a temporary decrease in the real rate (Brunner, Cukierman, & Meltzer, 1983), (d) through the price level (Bomhoff, 1982).

8. For example, more stimulation accompanied by higher inflation increases the welfare of the currently unemployed and decreases the welfare of retired individuals who hold fixed-interest financial assets. See also Hetzel (1984).

9. Our model does not give the government superior knowledge about the position of the economy. We allow the government, but not the public, to know the effect of current events on current policy objectives. The weights on the government's objectives may reflect the desires of the voters, but the public does not know the extent to which the government responds to voters in the current period.

be consistent with the actual policy strategy followed by government. The government's strategy is derived, in turn, by solving the maximization problem in equation (2), taking the process for the formation of $E[m_i|I_i]$ as given. In other words, $E[m_i|I_i]$ is a rational expectation of m_i formed by using the public's knowledge about the policymaker's strategy in conjunction with all the relevant information available. It is convenient to proceed in two steps. First we postulate the public's beliefs about the strategy that the government uses to set m^p. Then we show that when government optimizes (2), given the public's beliefs, the strategy that emerges is identical to the strategy that the public expects. The public's beliefs and the consequent form of the optimal predictor $E[m_i|I_i]$ are discussed in Section III. The solution of government's optimization in (2) and a proof of the rationality of the model are in Section IV.

III. The Public's Beliefs and the Expected Rate of Monetary Expansion

The public believes that the rate of monetary expansion planned by government is the following linear function of A and p_i

$$m_i^p = B_0 A + B p_i \qquad \text{for all } i \qquad (6)$$

where B_0 and B are known constants that ultimately depend on the underlying parameters of government's objective function. The public does not observe m_i^p directly. It observes

$$m_j = B_0 A + B p_j + \psi_j \qquad j \leq i - 1 \qquad (7)$$

which is a noisy indicator of m_j^p because of the existence of a control error. Since p_j displays a certain degree of persistence (as measured by ρ), past values of m are relevant for predicting the current rate of monetary growth. The information set of the public also contains the constants A, B_0, B, and ρ and the variances σ_v^2 and σ_ψ^2. As a consequence, from each past observation on m the public can, using (7), infer

$$y_j \equiv B p_j + \psi_j \qquad (8)$$

In Section IV we show that equation (7) is implied by the policymaker's actions given the public's belief, so this inference is correct for equilibrium positions.

It follows from this remark and equation (6) that

$$E[m_i|I_i] = B_0 A + B E[p_i|y_{i-1}, y_{i-2}, \ldots] \qquad (9)$$

It is shown in part 1 of the appendix that the conditional expected value on the right-hand side of (9) is

$$E[p_i|y_{i-1}, y_{i-2}, \ldots] = \frac{(\rho - \lambda)}{B} \sum_{j=0}^{\infty} \lambda^j y_{i-1-j} \tag{10a}$$

$$\lambda \equiv \frac{1}{2}\left[\frac{1+r}{\rho} + \rho\right] - \sqrt{\frac{1}{4}\left(\frac{1+r}{\rho} + \rho\right)^2 - 1} \tag{10b}$$

$$r \equiv B^2 \frac{\sigma_v^2}{\sigma_\psi^2} \tag{10c}$$

Substituting (10a) into (9), using (7) to express y_j in terms of m_j, and rearranging the resulting expression, we obtain

$$E[m_i|I_i] = \sum_{j=0}^{\infty} \lambda^j[(1-\rho)\bar{m}^p + (\rho - \lambda)m_{i-1-j}] \tag{11a}$$

$$\bar{m}^p \equiv B_0 A. \tag{11b}$$

Equations (6) and (11b) show that $\bar{m}^p$ is the mean and median (since m_i^p is normally distributed) value of planned monetary growth. The coefficient λ is bounded between zero and one. The optimal predictor of monetary growth is therefore a geometrically distributed lag, with decreasing weights, of weighted averages of the unconditional mean money growth and actually observed past rates of money growth. In general, individuals give some weight to mean governmental planned money growth and assign the rest of the weights to observations on actual past money growth.[10] It is easily checked that the sum of the weights on the mean and past rates of money growth is one.

The relative weight accorded to $\bar{m}^p$ is a measure of how strongly the public sticks to preconceptions rather than relying on actual developments. In the limit, as ρ tends to 1 the public abandons preconceptions entirely.[11] In this case, governmental preferences tend toward nonstationarity, so the information on the fixed mean $\bar{m}^p$ becomes of lesser significance. At the other extreme, when ρ tends to zero there is hardly any persistence in the stochastic component of governmental preferences, so the information on actual past rates of growth becomes less relevant for predicting the future. Hence individuals stick to preconceptions and give negligible weight to actual developments.[12]

---

10. Note that λ is always smaller than or equal to ρ so that the weight $\rho - \lambda$ is always nonnegative. This can be seen by noting from (10b) that the condition $\rho - \lambda \geq 0$ is implied by the condition $r \geq 0$, which is always the case.

11. Formally, when $\rho = 1$ the predictor reduces to Muth's (1960) predictor.

12. In the limit when $\rho \to 0$, $\lambda \to 0$ and the predictor tends to $\bar{m}^p$. However, we have not been able to show that the weight given to $\bar{m}^p$ is a decreasing function of ρ in all the range between zero and one, although such a result seems plausible.

IV. Derivation of Government's Decision Rule and Proof of the Rationality of Expectations

Policymakers choose the current planned rate of money growth using their objective function and knowledge of the current value of the random shock to this function. The history of policy constrains the effect of his choice on his objectives. In this section we first derive the policy strategy. Then we show that policymakers choose the strategy that the public expects.

Substituting (11b) into (11a), substituting the resulting expression into (3), substituting this result into equation (2), using (1), and rearranging, the policymaker's problem can be rewritten

$$\max_{\{m_i^p, i=0,1,\ldots\}} E_{G0} \sum_{i=0}^{\infty} \beta^i \left\{ \left[m_i^p + \psi_i - \frac{1-\rho}{1-\lambda} B_0 A \right. \right.$$
$$\left. \left. - (\rho - \lambda) \sum_{j=0}^{\infty} \lambda^j (m_{i-1-j}^p + \psi_{i-1-j}) \right] x_i - \frac{(m_i^p)^2}{2} \right\} \quad (12)$$

The policymaker chooses the actual value of m_0^p and a contingency plan for m_i^p, $i \geq 1$. Recognizing that in each period in the future the policymaker faces a problem that has the same structure as period's zero problem, the stochastic Euler equations necessary for an internal maximum of this problem are, following Sargent (1979, Chap. XIV),

$$x_i - (\rho - \lambda)\beta E_{Gi}[x_{i+1} + \beta\lambda x_{i+2} + (\beta\lambda)^2 x_{i+3} + \ldots] - m_i^p = 0 \quad i = 0, 1, \ldots \quad (13)$$

Equation (13) yields the actual choice of m_0^p and the contingency plan for all future rates of money growth for $i \geq 1$.[13]

Although the policymaker knows x_i in period i (and the public does not), he is uncertain about values of x beyond period i. On the basis of information available to him in period i, he computes a conditional expected value for x_{i+j}, $j \geq 1$. In view of (4) and (5) this expected value is

$$E_{Gi} x_{i+j} = A + E_{Gi} p_{i+j} = A + \rho^j p_i = \rho^j x_i + (1 - \rho^j) A \quad j \geq 0 \quad (14)$$

Substituting (14) into (13) using (4) and the formulas for infinite geometric

---

13. Note that the transversality condition

$$\lim \beta^i E_{G0}[x_i - (\rho - \lambda)\beta E_{Gi} \sum_{j=0}^{\infty} (\beta\lambda)^j x_{i+j+1} - m_i^p] = 0$$

is satisfied for any $\beta < 1$, since the term inside the brackets following E_{G0} is finite. This condition is sufficient for an internal maximum.

progressions, and rearranging:

$$m_i^p = \frac{1-\beta\rho}{1-\beta\lambda} A + \frac{1-\beta\rho^2}{1-\beta\rho\lambda} p_i \qquad (15)$$

Rationality of expectations implies that the coefficients of A and p_i should be the same across equations (15) and (6), respectively, so

$$B_0 = \frac{1-\beta\rho}{1-\beta\lambda(B)} \qquad (16a)$$

$$B = \frac{1-\beta\rho^2}{1-\beta\rho\lambda(B)} \qquad (16b)$$

The dependence of λ on B through equation (10) is stressed by writing λ as a function of B. Equation (16b) determines B uniquely as an implicit function of β, ρ, σ_v^2, and σ_ψ^2. Uniqueness is demonstrated by noting that the right-hand side of (16b) is increasing in λ and that λ is decreasing in r.[14] The definition of r in terms of B from equation (10c) implies therefore that both λ and the right-hand side of (16b) are monotonically decreasing in B. Hence this equation yields a unique solution for B. Given this solution, equation (16a) determines B_0.

As long as the optimal predictor of money growth is linear in the information set of the public, this solution is also unique. This can be shown by writing the optimal predictor as a general linear function of all past rates of money growth and by allowing m_i^p to be a general function

$$m_i^p = F(p_i, p_{i-1}, p_{i-2}, \ldots)$$

of the entire history of governmental objectives. Note that this formulation is similar to that of Green and Porter (1984) in which the current action of a representative firm depends on the entire history of the industry price.

The proof of uniqueness proceeds by substituting the general linear predictor into government's objective function in (2), deriving the Euler equations, and showing that they imply the function F above to be a linear function of p_i only, as postulated in equation (6). A proof is available on request.

Note that even if the costs of inflation to the policymaker had been expressed in terms of actual rather than planned money growth, the resulting equilibrium would be the same. The reason is that for any $i \geq 0$

$$(m_i)^2 = (m_i^p)^2 + \psi_i^2 + 2m_i^p \psi_i$$

The second term on the right-hand side of this equation does not depend on m_i^p and the third term has an expected value of zero, given the policymaker's

---

14. Let $b \equiv \rho + (1+r)/\rho$. Then from (10b), $\partial\lambda/\partial b = (1/2)\sqrt{\frac{b^2}{4}-1}\left(\sqrt{\frac{b^2}{4}-1}-b\right)$ which is negative since $(3/4)b^2+1 > 0$. Since b is increasing in r, λ is decreasing in r.

information in period 0. Hence the Euler equations for this reformulation are still given by (13), leading to the equilibrium solution in (15).

V. Credibility and the Determinants of the Distribution of Monetary Growth

Credibility is defined as the absolute value of the difference between the policymaker's plans and the public's beliefs about those plans. The smaller this difference, the higher the credibility of planned monetary policy. Credibility is relatively low when governmental objectives undergo large changes. In addition it is lower on average the longer it takes the public to recognize a change in governmental objectives. The weight λ in equation (10b) measures the degree of sluggishness in expectations. The higher the value of λ, the longer is the "memory" of the public and the less important are recent developments for the formation of current expectations. With a low λ, past policies are quickly forgotten. It can be shown that λ is a decreasing function of σ_v^2 and an increasing function of σ_ψ^2 [15]; the effects of past choices of monetary growth on current expectations are smaller in comparison with more recent choices the larger the value of σ_v^2 and the lower the value of σ_ψ^2. People give less weight to the more distant past the larger the variance of the innovation to governmental objectives and the lower the variance of the control error. The worse the control of the money stock (high σ_ψ^2), the longer will past policies affect future expectations.

Since λ is related to the speed with which the public recognizes shifts in governmental objectives, it is a prime determinant of the credibility accorded to new objectives of the government. Suppose that after remaining above its mean value m_i^p decreases below the mean. This more conservative attitude toward inflation will take longer to be recognized by the public the larger is λ. Therefore the worse the control of the money stock the lower the credibility of shifts to rates of monetary growth that differ from those previously experienced. λ can therefore be taken as a measure of credibility. The higher the value of λ, the longer it takes the public to recognize a change in governmental objectives and the lower, therefore, the government's credibility.

We turn now to the investigation of the distribution of money growth. Substituting (15) into (7), actual money growth can be rewritten

$$m_i = \frac{1 - \beta\rho}{1 - \beta\lambda} A + \frac{1 - \beta\rho^2}{1 - \beta\rho\lambda} p_i + \psi_i$$

15. Let $a \equiv \sigma_v^2/\sigma_\psi^2$. The total effect of a change in a on λ, taking the dependence of B on a through (16b) into consideration, is

$$\frac{d\lambda}{da} = \frac{\partial \lambda}{\partial b}\left[\frac{B^2(1 - \beta\rho\lambda)}{\rho(1 - \beta\rho\lambda - 2a\beta\rho B^2 \partial\lambda/\partial r)}\right]$$

which is negative since $\partial\lambda/\partial b$ and $\partial\lambda/\partial r$ are both negative as implied by footnote 14. The result in the text follows by noting that a is positively related to σ_v^2 and negatively related to σ_ψ^2.

with unconditional mean and variance that are given respectively by

$$Em_i = \frac{1-\beta\rho}{1-\beta\lambda} A \tag{17}$$

$$V(m_i) = \left[\frac{1-\beta\rho^2}{1-\beta\rho\lambda}\right]^2 \frac{\sigma_v^2}{1-\rho^2} + \sigma_\psi^2 \equiv B^2 \frac{\sigma_v^2}{1-\rho^2} + \sigma_\psi^2 \tag{18}$$

Since $A > 0$, $\rho < 1$, and $\lambda \leqslant \rho$, the average rate of monetary expansion is always positive. Equation (17) suggests that mean monetary expansion is systematically related to the underlying parameters of the model. The following proposition summarizes the effects of some of those parameters on average monetary growth.

Proposition 1: Monetary growth is larger (a) higher A and (b) the higher σ_ψ^2.

Part (a) follows immediately from (17) by noting that $\lambda \leqslant \rho < 1$. Part (b) follows by noting that Em_i is an increasing function of λ, which is in turn an increasing function of σ_ψ^2 (footnote 15).

Part (a) of proposition 1 says that average monetary growth is higher when the policymaker is more biased toward economic stimulation than to preventing inflation (high A). Part (b) implies that average monetary growth is also higher the less effective is control of monetary growth as measured by a relatively high σ_ψ^2. The reason is that at higher σ_ψ^2 it takes longer for the public to recognize a shift to a more expansionary policy. The negative effects of increased monetary expansion on government's objectives are delayed, so the government gains more from trading future inflation for current economic stimulation.

The following proposition summarizes the effects of the underlying parameters of the model on the variability of monetary growth.

Proposition 2: (a) For any $\beta < 1$, the variance of monetary growth is larger the larger the σ_ψ^2. (b) For a given finite value of the variance of the monetary control error, σ_ψ^2, the variance of monetary growth is larger the lower the discount factor β.

Part (a) follows from (18) by noting that $V(m_i)$ is an increasing function of σ_ψ^2 both directly and through its dependence on λ. Part (b) is proved by noting that $V(m_i)$ is increasing in B and that B is decreasing in β.[16]

Variability and uncertainty are not identical (Cukierman, 1984, Chap. 4, Sect. 4). A natural measure of monetary uncertainty is the variance of the money growth forecast error

$$V(e) = E[m - E[m_i|I_i]]^2$$

---

16. The last relation follows by differentiation of (16b) with respect to β. This yields $\partial B/\partial \beta = 2B\rho(\lambda - \rho)/(1 - \beta\rho\lambda)^2$. This expression is nonpositive, since $\lambda \leqslant \rho$, and is strictly negative for a finite σ_ψ^2.

Proposition 3: For a given value of the variance of the monetary control error σ_ψ^2, monetary uncertainty as measured by $V(e)$ is larger the lower the discount factor β. The proof is developed in part 2 of the appendix.

Part (a) of proposition 2 states that the variance of monetary growth is larger the larger the variance of the control error in money growth. There are two reasons. First is the direct effect. For any level of planned growth actual monetary growth is more variable. Second is the effect on λ. The public is slower to detect shifts in governmental objectives, so it pays government to induce a higher degree of stimulation by creating more uncertainty. For a similar reason, when the discount factor is low, government discounts the costs associated with expectations of future inflation more heavily, and chooses more current stimulus. As suggested by proposition 3, this is done by creating more uncertainty, part of which takes the form of higher variability.

VI. Comparison with Previous and Current Literature

The deterministic component of the policymaker's objective function in (2) is similar to that used by Barro and Gordon (1983b). As a result, our solution exhibits a similar inflationary bias that increases with A, the average relative preference of the policymaker for economic stimulation (proposition 1a). A novel element of our framework is that the public's information about the shifting objectives of the policymaker changes over time.[17] Shifting objectives and noisy control permit people to supplement the information on average objectives, given by A, by observing past money growth. Our solution specializes to Barro and Gordon's discretionary solution when asymmetric information is removed from the model. Formally, asymmetric information is eliminated when for a given (nonzero) σ_ψ^2 the variability of governmental objectives, σ_v^2, is zero. In this case the actual relative preference of the policymaker for stimulation is common knowledge and is given by A. When $\sigma_v^2 = 0, r = 0$, which implies through (10b) that $\lambda = \rho$[18] and through (4) and (5) that $\rho_i \equiv 0$ for all i. In this case the solution in (15) specializes to

$$m_i^p = A \quad \text{for all } i \geq 0$$

which is the solution obtained by Barro and Gordon under discretion. In addition it can be seen from (9) that in this case

$$E[m_i | I_i] = A$$

---

17. Barro and Gordon (1983a, 1983b) limited their analysis of discretionary policy to the case in which the public does not need to learn about changes in the policymaker's objectives because those objectives are fixed.

18. For $r = 0$, (10b) implies that $\lambda = (1/2)(1/\rho + \rho) - \sqrt{(1/4)(1/\rho + \rho)^2 - 1}$, which is equal to ρ.

That is, as in Barro and Gordon (1983b, p. 595), individuals do not pay attention to actual rates of money growth in forming expectations. Since they know with certainty the structure of governmental objectives, they do not need to use information about observed rates of money growth to forecast future growth. Whatever the realization of money growth, individuals interpret its deviation from A as a transitory control deviation and stick to the preconceived notion that future monetary growth will be A. In this particular case the only way to change expectations is by convincing the public that discretion has been abandoned in favor of a different regime.

In the more general case considered here, expectations depend both on the preconception $\bar{m}^p$ in (11a) and on the history of money growth. This last dependence is induced by asymmetric information between the policymaker and the public. Correspondingly, as can be seen from (15), planned monetary growth is composed of two components. The first, which depends on A, is common knowledge. The second, which depends on p_i, is known with certainty only to the policymaker. It is this second component that makes expectations dependent on past rates of money growth and makes it possible to change expectations without necessarily changing the discretionary nature of the policy regime. Note that the relative weights given to the preconception $\bar{m}^p$ and the history of monetary growth depend on the degree of persistence, ρ, in that part of governmental objectives about which the policymaker possesses an information advantage. When ρ tends to one, the weight given to the preconception $\bar{m}^p$ becomes negligible (see (11a)).

In Barro and Gordon (1983b) and in the example introduced by Kydland and Prescott (1977, pp. 477–80), a rule that binds the policymaker to set $m_i^p = 0$ in all periods is believed, so the expectation of money growth is zero. The rule achieves a better value for the policymaker's objective function than does discretion. None of these frameworks, however, incorporates asymmetric information. In the presence of asymmetric information a zero rate of money growth rule does not always achieve a better value for the policymaker's objective function. The reason is that the public is slow to recognize shifts in governmental objectives. As a result, for sufficiently unstable objectives and slow adjustment of expectations, the positive contribution of a current positive e_i to the policymaker's objective function in (2) can dominate the negative effects of higher inflation and future negative values of e_i by enough to make discretion preferable to a binding zero rate of money growth rule.[19] In such cases there is no difference between the discretionary-consistent solution and the optimal solution.

In their reputational paper, Barro and Gordon (1983a) postulate an exogenously given "punishment" period.[20] They also briefly consider an extension of their basic model in which government has an information advantage. However, even in the extension the expected rate of monetary

---

19. An example for which this is the case is discussed at the end of Section VII.

20. This term is borrowed from the repeated games literature to describe the relationship between past choices of monetary growth and current expectations. In this literature, the public chooses its expectation so as to induce socially desirable behavior by government in a supergame.

growth is fixed and independent of actual changes in monetary growth. This expectation is rational, given the stochastic structure postulated by Barro and Gordon. But their structure lacks descriptive realism, since forecasts of inflation are usually influenced by actual inflation and monetary growth. A dramatic example is the decrease in expected inflation between the end of the 1970s and the present. Further, recent empirical work leaves no doubt that expectations regarding future monetary growth and inflation change within a discretionary framework (see, inter alia, Hardouvelis, 1984, and the many references there). Explicit modeling of the way expectations change is essential for discussing changes in credibility in the absence of a constitutional rule.

The present framework links expectations to observed money growth by introducing persistence in the policymaker's objectives and imperfect monetary control. This permits us to discuss different degrees of credibility within a discretionary regime. The evidence presented in Hardouvelis (1984) suggests that in spite of the fact that U.S. monetary policy has remained discretionary, the 1979 change in the Fed's operating procedures changed the public's perception of the objectives of the Fed. The model of Barro and Gordon does not handle a change in perceptions of this type.

An additional advantage of our formulation is that it relates the speed with which expectations adjust to the quality of monetary control and other parameters of the environment. The determinants of the size and timing of "punishment" are identified rather than postulated exogenously as in Barro and Gordon. Further, as stressed by Backus and Driffill (1985), a weakness of the Barro and Gordon analysis is that their equilibrium solution depends critically on the form this punishment strategy takes. As a result their model has multiple equilibria with no mechanism for choosing among them. We have shown that, at least within the class of linear minimum-square-error predictors of money growth, our equilibrium is unique. Given linearity, the attempts by individuals to minimize their error of forecast determines a unique pattern of learning that acts as a deterrent to the policymaker. This differs conceptually from Barro and Gordon's (1983a) punishment mechanism. In Barro and Gordon, the public is viewed as a single player who picks his expectation strategically in order to induce "better" behavior on the part of the policymaker. Here the structure of deterrence is induced, as a by-product of each individual's attempt to minimize his forecast error.

As stressed by Bull (1982), casual empiricism suggests that a positive amount of resources is devoted to monitoring the central bank. Our framework is consistent with this observation; it is rational to study central bank behavior. Further, because expectations are influenced by past monetary growth, our concept of discretionary policy permits the policymaker to consider the effects of current policy on future expectations. In contrast, Barro and Gordon (1983a, p. 106) restrict discretion to mean that "... the policymaker treats the current inflationary expectation, and all future expectations, as given when choosing the current inflation rate." Actual policymaking in the absence of a constitutional rule recognizes the effect of current policy actions on future expectations, perhaps with some discounting. Barro and Gordon's definition of discretion seems overly restrictive, since it applies only to a world in which current policy actions do not affect future expectations.[21]

The wider notion of discretion we use applies to any arrangement in which there is no constitutional rule to restrict the range of possible actions of the policymaker. The policymaker may follow a decision rule, in the sense of the optimal control literature, that takes account of the effect of present policy actions on future expectations. The use of a decision rule permits the policymaker to maximize his objective function. The policymaker, in our analysis, is free to choose the weights he places on the arguments of the decision rule. The contingent decision rule that we derive corresponds to discretionary policy, since no a priori restrictions are imposed on the feasible set of policy actions. In contrast, a Friedman-type rule in which policy actions are subject to a priori binding constraints is a constitutional rule.

Recently Backus and Driffill (1985) have formulated credibility as the outcome of a sequential equilibrium in a repeated game. Our model shares with theirs the asymmetric information about government objectives and the notion that discretionary policy is dynamically consistent. It differs in that government objectives are allowed to change continuously through time and to assume an infinite number of values. In Backus and Driffill there are only two possible types of government, and these types never change. Consequently, inflationary expectations can assume only two possible values. Moreover, once a government inflates it is revealed to be the weak type, and asymmetric information is eliminated forever. Backus and Drifill restrict government's discount factor to 1 and do not analyze the effect of the precision of monetary control on the choice of policies. They define credibility in terms of the probability that government is "hard nosed" (has no incentive to inflate), whereas we conceive of credibility as the speed with which the public detects changes in the policymaker's objectives.

As in Backus and Driffill, imperfect credibility arises here without any dynamic inconsistency of the type defined by Kydland and Prescott (1977, p. 475). The reason dynamic inconsistency does not arise in our framework is that the "action" taken by the public—evaluating $E[m_i|I_i]$—does not depend on the future settings, $m^p_{i+j} (j \geq 1)$, of government's choice of instruments (see (11)). In cases of this kind, Kydland and Prescott (1977, p. 476) point out that the time-consistent solution is also optimal. When period $i + j (j \geq 1)$ arrives, the government follows the contingency plan made in period i. At the risk of repetition it should be pointed out that this is not surprising, since our solution corresponds conceptually to Kydland and Prescott's discretionary consistent solution in their example on pages 477 through 480.

21. Our framework differs from that of Barro and Gordon in several other respects. Our objective function represents the attempt of the policymaker to elicit support for his policies, while Barro and Gordon interpret the objective function of the policymaker as a social welfare function. Imperfect credibility in Barro and Gordon concerns the socially optimal constitutional rule and occurs because the policy rule is not dynamically consistent. In our framework credibility is imperfect even with a dynamically consistent discretionary policy because of noisy control and shifting objectives. Taylor (1983) raise doubts about the relevance of the Barro-Gordon model as a positive theory of inflation. Canzoneri (1985) tries to resolve the doubts by appealing to asymmetric information.

The basis for imperfect credibility here is the policymaker's advantage over the public, which is due to shifting objectives and noisy control. Policymakers know their stochastically changing objectives, but the public does not. The best the public can do is to form expectations, allowing for this noise, and use all the information available each period to infer current and future money growth.

VII. The Politically Optimal Level of Ambiguity

To this point we have considered the level of noise in the control of money as a technologically given parameter. Suppose, however, that technology only puts a lower bound $\underline{\sigma}_\psi^2$ on the variance of the control error. The policymaker can choose any $\sigma_\psi^2 \geq \underline{\sigma}_\psi^2$. We assume for simplicity that $\underline{\sigma}_\psi^2 = 0$.

In this section the policymaker sets the value of σ_ψ^2 once and for all so as to maximize the long-run expected value of his objective function in (12). The choice of this variance determines the politically optimal level of ambiguity in the conduct of monetary policy, since a higher choice of σ_ψ^2 conveys more ambiguous signals to the public.

For a given level of σ_ψ^2 the optimized value of the objective function of the government is obtained by substituting the optimal choice of monetary growth (equation (15)) into the obtective function of the government in (12). Abstracting from the conditional expected-value operator in (12) yields

$$J(\{\psi_i\}_{-\infty}^\infty, \{p_i\}_{-\infty}^\infty) \equiv \sum_{i=0}^\infty \beta^i \left[\left\{ B_0 A + B p_i + \psi_i - \frac{1-\rho}{1-\lambda} B_0 A - (\rho - \lambda) \right. \right.$$
$$\left. \left. \sum_{j=0}^\infty \lambda^j (B_0 A + B p_{i-1-j} + \psi_{i-1-j}) \right\} x_i - (m_i^p)^2 \right] \quad (19)$$

where $\{\psi_i\}_{-\infty}^\infty$ and $\{p_i\}_{-\infty}^\infty$ denote the sequences of ψ and p.

Since the policymaker sets σ_ψ^2 on the basis of the long-run value of his objective function rather than on the basis of particular recent realizations of x_i, the relevant objective function for the choice of σ_ψ^2 is the unconditional expected value of $J(\cdot)$. It is shown in part 3 of the appendix that this expected value is

$$G'(\sigma_\psi^2) \equiv EJ(\cdot) = \frac{\sigma_v^2}{1-\beta} \left[\frac{1-\beta\rho^2}{(1-\beta\rho\lambda)(1-\rho\lambda)} - \frac{(1-\beta\rho^2)^2}{2(1-\beta\rho\lambda)^2(1-\rho^2)} \right.$$
$$\left. - K^2 \frac{(1-\beta\rho)^2}{2(1-\beta\lambda)^2} \right] \quad \text{where } K \equiv \frac{A}{\sqrt{\sigma_v^2/(1-\rho^2)}} = \frac{A}{\sigma_p} \quad (20)$$

The first term in the brackets on the right-hand side of (20) represents the mean (positive) contribution of economic stimulation to governmental objectives. The mean value of economic stimulation through surprise creation is zero, since negative and positive surprises cancel each other on average. But the contribution of monetary surprises to governmental objectives is positive on average. The reason is that the rate of money growth is positively related to the marginal benefit of surprise creation to the government. As can be seen from (15), when the marginal benefit of a surprise is higher than average ($x_i > A \leftrightarrow p_i > 0$) government chooses a higher than average rate of monetary growth, and when the marginal benefit of a surprise is lower than average ($x_i < A \leftrightarrow p_i < 0$) the government chooses a lower than average rate of monetary growth. Consequently, when government cares more than on average about economic stimulation, surprises are positive on average, and when it cares less than on average about economic stimulation, surprises are negative on average, making the unconditional expected value of the benefits from surprise creation positive. The government derives a positive gain, on average, from the ability to create surprises because it can allocate large positive surprises to periods in which x_i is relatively high and can leave the inevitable negative surprises for periods with relatively low values of x_i. More formally, by using (7) and (11a), it can be shown that

$$E \sum_{i=0}^{\infty} \beta^i e_i x_i = \frac{B}{1-\beta} Ep_i(p_i - p_i^*) \tag{21}$$

where

$$p_i^* = (\rho - \lambda) \sum_{j=0}^{\infty} \lambda^j p_{i-1-j}$$

Obviously $E(p_i - p_i^*) = 0$. But $Ep_i(p_i - p_i^*) > 0$ since there is a positive correlation between p_i and $p_i - p_i^*$. This positive correlation is created by the government's attempt to maximize its objective function, which leads to the contingent behavior summarized in (15). By contrast, when the government does not possess an information advantage, $\sigma_v^2 = 0$ and p_i is identically zero. This implies that the expression in (21) is zero as well. Thus the existence of asymmetric information makes it possible for government to attain a higher value for its objectives through surprise creation in a time-consistent equilibrium.

The last two terms on the right-hand side of (20) represent the mean (negative) contribution of inflation to governmental objectives. The effect of σ_ψ^2 on $G(\cdot)$ comes from its effect on λ. Since λ is monotonically increasing in σ_ψ^2 and all the other elements that affect λ are fixed parameters, the choice of σ_ψ^2 is equivalent to a choice of λ. Hence the choice of the tightness of control procedures by the government can be expressed formally as

$$\max_{\lambda} G(\lambda) \equiv \max_{\sigma_\psi^2} G'(\sigma_\psi^2) \tag{22}$$

where $G'(\cdot)$ is given in (20).

If the policymaker chooses perfect control, $\sigma_\psi^2 \to 0$, and from (10b) $\lambda \to 0$. At the other extreme, when $\sigma_\psi^2 \to \infty$, $\lambda \to \rho$ from below.[22] Hence the range of choice open for λ is from a minimum of zero (which corresponds to perfect control) to a maximum of ρ (which corresponds to the minimum possible amount of control).

A sufficient condition for a positive (politically) optimal level of ambiguity ($\sigma_\psi^2 > 0$) is that $G(\lambda)$ be an increasing function of λ at $\lambda = 0$. Using (20) and differentiating (22) with respect to λ,

$$F(\lambda) \equiv \frac{\partial G}{\partial \lambda} = \frac{\sigma_v^2}{1-\beta}\left[\frac{\rho(1-\beta\rho^2)(1+\beta-2\beta\rho\lambda)}{(1-\beta\rho\lambda)^2(1-\rho\lambda)^2} - \frac{\beta\rho(1-\beta\rho^2)^2}{(1-\rho^2)(1-\beta\rho\lambda)^3} \right.$$
$$\left. - \beta K^2 \frac{(1-\beta\rho)^2}{(1-\rho^2)(1-\beta\lambda)^3}\right] \quad (23)$$

By manipulating the monetary control parameter, σ_ψ^2, the government affects the speed with which the public becomes aware of changes in governmental objectives and therefore the average value of benefits from surprise creation. In particular, an increase in ambiguity increases the mean value of benefits from surprise creation[23] but also increases the average rate of inflation, which is bad from government's point of view. The level of monetary ambiguity is chosen by weighting the effects of those two conflicting elements on governmental objectives.

At $\lambda = 0$, the expression in (23) reduces to

$$F(0) \equiv \frac{\sigma_v^2}{1-\beta}\left\{\rho(1-\beta\rho^2)(1+\beta) - \beta\left[\frac{\rho(1-\beta\rho^2)^2}{1-\rho^2} + K^2\frac{(1-\beta\rho)^2}{1-\rho^2}\right]\right\} \quad (24)$$

The expression in (24) is positive if $\rho > 0$ and β is sufficiently small. Hence a strong degree of time preference on the part of government leads to a choice of institutions that produce loose control of the money supply, even if perfect control is technologically feasible. For a sufficiently high value of β, G is decreasing at $\lambda = 0$, so the politically optimal level of control *may* be perfect. In other words, a negative value of (24) (which obtains for sufficiently high β) is a necessary but not sufficient condition for the optimal value of σ_ψ^2 to be zero.

At the other extreme, when $\lambda = \rho$, (23) becomes

$$F(\rho) \equiv \frac{\sigma_v^2}{(1-\beta)(1-\rho^2)}\left[\frac{\rho(1+\beta-2\beta\rho^2)}{(1-\beta\rho^2)(1-\rho^2)} - \beta\left(\frac{\rho}{1-\beta\rho^2} + K^2\frac{1}{1-\beta\rho}\right)\right]$$

This expression is negative for a sufficiently large β and ρ bounded away from one. For this case a sufficiently low degree of time preference on the part of government is necessary for a finite optimal level of ambiguity, $\sigma_\varepsilon^2 < \infty$.

22. For $\sigma_\psi^2 \to \infty$, $r \to 0$, which implies $\lambda \to \rho$. See also footnote 18.
23. That is, the first term inside the brackets on the right-hand side of (23) is positive, since $1 + \beta - 2\beta\rho\lambda > 0$. The intuition again is that in periods with high values of x_i, the positive effect of postponing the adjustment of expectations on governmental objectives is larger than the negative effect of such a postponement when x_i is below its mean value.

Let λ^* be the value of λ that maximizes the expression in equation (22). The previous discussion suggests that λ^* could be equal to ρ, zero, or some intermediate value between ρ and zero. Correspondingly the politically optimal value of σ_ψ^2 is infinite, zero, or some positive but finite value. For all $\beta < 1$ and $\sigma_v^2 > 0$, equation (23) implies that necessary conditions for $\lambda^* = \rho$, $0 < \lambda^* < \rho$, and $\lambda^* = 0$ are $F(\rho) \geq 0$, $F(\lambda^*) = 0$, and $F(0) \leq 0$, respectively. The following proposition provides a characterization of the three types of solution in terms of the parameters (ρ, β, and K^2) by providing a sufficient condition for each type of solution.

Proposition 4: Let

$$c_1 \equiv \frac{\rho}{\beta} \frac{1}{(1-\rho\beta)^2} \left[\frac{1}{1-\rho^2} + \frac{\beta}{1-\rho^2\beta} - \beta(1-\rho^2\beta)^2 \right]$$

$$c_H \equiv \frac{\rho}{\beta} \frac{1-\rho^2\beta}{(1-\rho\beta)^2} \{1 - \rho^2[1+\beta(1-\beta)]\}$$

$$c_L \equiv \frac{\rho}{\beta} \frac{1-\rho\beta}{1-\rho^2}$$

$$c_2 \equiv \frac{\rho}{\beta} \frac{1-\rho\beta}{1-\rho^2\beta} [(1-\rho^2)(1-\rho^2\beta)(1+\beta) - \beta]$$

then for $K^2 \equiv A^2/\sigma_p^2$,

(i) $c_2 \leq c_1$.
(ii) If $K^2 < c_2$, $\lambda^* = \rho$.
(iii) If $K^2 > c_1$, $\lambda^* = 0$.
(iv) If $c_L < K^2 < c_H$, $0 < \lambda^* < \rho$ and $c_2 \leq c_i \leq c_1$ for $i = L, H$.

Proof: See part 4 of the appendix.

Proposition 4 states that if K^2 is larger than c_1, the optimal level of ambiguity is zero. If K^2 is smaller than c_2, the optimal level of ambiguity is as large as possible given the underlying stochastic structure. Finally, if $c_L < K^2 < c_H$, there is an internal solution for the optimal level of ambiguity. Figure 8-1 illustrates the different ranges of K^2 and the corresponding solution types for λ^*. The range (c_H, c_L) may be empty depending on the values of the discount factor β and the degree of persistence in governmental objectives as measured by ρ. It is more likely to be nonempty for lower values of ρ and for higher values of β.[24]

Proposition 4 and Figure 8-1 suggest that for given ρ and β the size of K is crucial for the determination of the solution type. K^2 increases directly with A^2 and inversely with σ_p^2. The latter is a measure of the uncertainty about the

---

24. However, even when c_H and c_L are such that $c_H < c_L$ so that condition (iv) of proposition 4 is not satisfied, λ^* *may* be internal if $c_2 < K^2 < c_1$. Condition (iv) is sufficient but not necessary for an internal solution.

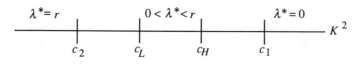

Figure 8-1.

government's objectives. The larger the uncertain part in governmental objectives in relation to the certain part A, the smaller is K and the more likely that ambiguity is high. At the other extreme, when K is very large, say $K \to \infty$, there is almost no uncertainty with respect to governmental objectives.[25]

Proposition 4 suggests that there is a monotonic relation between instability in government's objectives and the optimal level of ambiguity. When the level of instability in governmental objectives is high, so that $K^2 < c_2$, the optimal level of ambiguity is high too. When the level of instability in governmental objectives is relatively low, so that $K^2 > c_1$, the optimal level of ambiguity is zero. This monotonic relationship between the level of instability in governmental objectives and the degree of ambiguity embodied in monetary institutions holds also for internal solutions. Proposition 5 summarizes the effect of larger uncertainty in relative governmental objectives on the precision of monetary institutions chosen by government.

Proposition 5: For given values of β and ρ, (i) if $0 < \lambda^* < \rho$, an increase in K^2 decreases λ^*, and (ii) if λ^* is at a corner (0 or ρ), an increase in K^2 does not increase λ^*.

Proof: When $0 < \lambda^* < \rho$, $F(\lambda^*) = 0$. Using this fact, (23), and the implicit function theorem, we obtain

$$\frac{d\lambda^*}{dK^2} = \frac{1}{\partial^2 G/\partial \lambda^2} \left[\frac{\beta(1-\beta\rho)^2}{(1-\rho^2)(1-\beta\lambda^*)^3} \right]$$

This derivative is negative, since $\partial^2 G/\partial \lambda^2 < 0$ by the second-order condition for an internal maximum and the fact that β, ρ, and λ are all bounded between zero and one. This establishes (i).

When $\lambda^* = \rho$, an increase in K^2 *may* decrease λ^* but will not increase it, since λ^* is already at its maximal feasible value. When $\lambda^* = 0$, this implies that even if $F(\lambda)$ is positive for some λ's in the range $0 < \lambda < \rho$, it is not sufficiently positive since $\lambda^* = 0$. Equation (23) implies that an increase in K^2 decreases $F(\lambda)$ more the higher the value of λ. Hence if, for the original K^2, $\lambda^* = 0$ it will, a fortiori, be zero for a larger K^2.

Proposition 5 has a useful, intuitive meaning. When governmental objectives are relatively unstable, it pays a rational public to give a lot of weight to recent developments in forecasting of the future rate of growth of money. For a given

---

25. For a given nonzero A this is equivalent to the case $\sigma_v^2 = 0$ (and $\sigma_p^2 = 0$) discussed in Section VI.

quality of monetary control (a given σ_ψ^2), individuals are more sensitive to recent developments in an economy with relatively unstable governmental objectives, making it more difficult for government to exploit the benefits of monetary surprises. By increasing σ_ψ^2, the government with relatively unstable objectives can partially offset this effect by increasing the length of time it takes the public to detect a given shift in its objectives.

The same effect can be seen slightly differently by noting that, ceteris paribus, the larger the variance of the innovation to governmental objectives, σ_v^2, the higher the optimal level of σ_ψ^2. For given values of ρ, β, and K^2 there is an optimal level of credibility that is represented by the learning parameter λ^*. This parameter induces through equation (10b) an optimal level of r that is denoted r^*. Given r^*, the value of σ_ψ^2 chosen by government is larger, or at least not smaller, the larger the σ_v^2.[26] This leads to the following proposition.

Proposition 6: For given values of ρ, β, and K^2, the politically optimal level of σ_ψ^2 is a nondecreasing function of σ_v^2.

Equation (23) implies that two countries with identical values of the parameters ρ, β, and K^2 will have identical levels of credibility, since λ^* is the same in both. But proposition 6 implies that the country with the higher σ_v^2 will achieve this common credibility level by choosing a higher level of noise in the control of money.

A large body of empirical evidence (Jaffe & Kleiman, 1977; Logue & Willet, 1976; Okun, 1971) suggests that the level and the variability of inflation are positively related across countries. Our framework links this positive relationship to differences in the relative instability of governmental objectives across countries. More precisely, for a given A, a country with a higher σ_p^2 will have a lower K^2 and, by proposition 5, a higher σ_ψ^2. Propositions 1 and 2 imply that, ceteris paribus, a higher σ_ψ^2 implies both higher average monetary growth and higher monetary variability.[27] Hence if political instability varies across countries, this variation will produce a positive relation between average money growth and the variability of money growth. This, in turn, induces a positive relationship between the average level and the variability of inflation across countries.

We now consider the effect of the policymaker's discount factor β on the optimal level of ambiguity. Since for a given σ_v^2, ρ, and A the optimal level of credibility, λ^*, and the optimal degree of ambiguity as embodied in σ_ψ^2 are monotonically related, it is enough to find the effect of β on λ^*. For internal solutions the direction of this effect depends on the sign of the partial derivative of $F(\lambda^*)$ (from (23)) with respect to β. The sign is ambiguous in general. For corner solutions at 0 or ρ, the sign depends on the effect of β on $F(\lambda)$ in the range

---

26. The reason is that an increase in $a \equiv \sigma_v^2/\sigma_\psi^2$, taking into account the resulting change in B, causes an increase in r. This follows from the fact that $dr/da = B(B + 2aB^3\beta\rho(\partial\lambda/\partial r)/(1 - \beta\rho\lambda - 2aB^2\beta\rho(\partial\lambda/\partial r)]$ is nonnegative, since $1 - \beta\rho\lambda \geq 0$. Hence the larger σ_v^2 the larger the value of σ_ψ^2 needed to attain the optimal values of r^* and of λ^*. This interpretation was suggested by Robert Hetzel.

27. $V(m_i)$ is larger also because of the direct effect of σ_p^2. See equation (18).

$0 \leq \lambda \leq \rho$. This sign is also ambiguous. In spite of the ambiguities, however, numerical simulations for a wide range of values of ρ, β, and K^2 suggest that in more than 99 percent of the cases examined λ^* is weakly decreasing in β.[28] Moreover, the few cases in which λ^* turned out to be strictly increasing in β, all occurred for $\beta > 0.8$. We conclude on the basis of these simulations that, except for a relatively small subset of parameters, the optimal level of ambiguity increases, or at least does not decrease, when the degree of time preference increases (β declines). Policymakers with short horizons and high time preference are likely to prefer more ambiguity and are therefore less credible.

We conclude this section by demonstrating that there are cases in which the unconditional expected value of government's objective function is larger under discretion than under a Friedman (1960)-type money growth rule that constrains the policymaker to set m_i^p, for all i, at zero. With a Friedman rule the policymaker does not have the ability to choose the timing of surprises. To the extent that surprises occur they are due to imperfect monetary control and are completely unrelated to changes in the objectives of the policymaker. As a consequence, the unconditional expected value of government's objective function is zero in this case.

Under discretion with $\lambda = \rho$[29], the unconditional expected value of governmental objectives from equation (20) reduces to

$$\frac{\sigma_v^2}{2(1-\beta)(1-\rho^2)} (1 - K^2) \tag{25}$$

which is positive for all $K < 1$. Hence if governmental objectives are sufficiently unstable the greater "natural" ability of the government to create surprises at the politically right time yields a better value to government's objectives under discretion than under a Friedman rule. This example, chosen mostly for its simplicity, illustrates a fundamental difference between symmetric information and asymmetric information with persistence in objectives. It relates directly to claims made in Section V (see footnote 19).

To sum up the main finding of this section, the politically optimal level of control over monetary growth is not necessarily the minimum level. It may be in the government's interest to pick institutions with loose control over the money stock even if better control is technologically feasible.[30]

---

28. Altogether 19,796 different combinations of the parameters ρ, β, and K^2 were examined. For each combination, λ^* was evaluated by numerical search. The combinations were formed from 7 values of ρ between 0 and 1, 28 values of K^2 between 0 and 10, and 101 equally spaced values of β in the [0, 1] interval. λ^* turned out to be increasing in β in only 117 cases.

29. The optimal value of λ may be lower than ρ. But if it is, the expected value of government's objectives will be at least as large as in equation (25) and therefore still better than the value achieved under a Friedman rule.

30. If the policymaker's objectives are a function of actual money growth (as suggested at the end of Section III), the incentive for ambiguity would be smaller, since a higher σ_ψ^2 increases the cost of inflation by increasing the unconditional expected value of m_i^2.

VIII. Concluding Comments

This chapter develops a politically based theory of credibility, monetary growth, and ambiguity for a monetary system in which control is imperfect and policymakers have an information advantage about their own objectives. For a technologically determined level of control, the credibility of newly instituted disinflationary policies depends on the quality of monetary control. With tight control, a few periods of determined slowdown in the rate of monetary expansion suffice to convince the public that money growth is permanently lower. As a consequence, expectations of inflation fall quickly. Unexpected rates of monetary growth remain negative for a relatively brief period, and the accompanying unemployment is relatively small. In this case a "cold turkey" disinflationary policy is preferable to "gradualism," since a large decrease in monetary expansion generates credibility relatively quickly. If the policymaker has poor control of money growth, disinflationary policy takes substantially longer to become credible, however. The interim period of unemployment is longer and unemployment is higher. The costs of disinflation are also higher. A gradual approach that permits the public to adjust anticipations seems preferable in these circumstances.[31]

Taylor (1980b) and Meyer and Webster (1982) analyze the response of inflation when learning is gradual. Using a nominal-contracts framework, Fischer (1984) provide estimates of the costs of disinflation for two alternative assumptions about the reaction of expectations to changes in policy. In one case there is no change in expectations regarding money growth, and in the other expectations are adaptive. His analysis suggests that the costs of disinflation are quite sensitive to the way expectations adjust. This chapter relates this critical speed of adjustment to some underlying factors such as the quality of monetary control and the degree of instability in the objectives of the policymaker. Fischer's analysis in conjunction with ours creates a link between the costs of disinflation on the one hand and the quality of monetary control and the degree of instability in the objectives of the policymaker on the other.

A main result of this chapter is that policymakers do not necessarily choose the most efficient control procedure available. Instead they may choose to increase ambiguity. We find that there is a politically optimal level of ambiguity.

Our finding that the government may prefer a higher to a lower level of ambiguity suggests an explanation for the Federal Reserve's inclination for secrecy, documented in Goodfriend (1986). The intuitive reason is that a certain degree of ambiguity provides policymakers with greater control of the timing of monetary surprises. When there is ambiguity about policy, they can create large positive surprises when they care most about stimulation and leave the inevitable negative surprises for periods in which they are relatively more concerned about inflation.

---

31. In Cukierman and Meltzer (1986), reprinted as Chapter 9, the analysis is extended to the case of mandatory announcements of monetary targets.

The policymaker determines the level of ambiguity by choosing the quality of monetary control. This choice determines, in turn, the speed with which individuals become convinced that the policymaker's objectives have changed when this is the case. Credibility depends on the speed with which the public learns; actions that delay learning lower credibility. We show that policymakers with relatively unstable objectives tend to be more ambiguous and less credible. That both the mean and the variance of monetary growth are positively related to the level of noise in monetary control implies the existence of a positive cross-sectional relationship between the mean and the variance of inflation. The existence of such a relation is widely documented (Jaffe & Kleiman, 1977; Logue & Willet, 1976; Okun, 1971).

The level of ambiguity chosen by policymakers is also systematically related to their degree of time preference. Our results suggest that more often than not policymakers with stronger time preference choose less precise control procedures. In addition, for a given precision of control procedures, monetary uncertainty is larger the larger is the degree of time preference of the policymaker.

Although we did not submit the theory to empirical verification, we believe it is consistent with a number of additional observations. First, rates of monetary growth vary considerably both over time within a given country and between countries. Second, the actual conduct of monetary policy in many countries, including the United States, corresponds more to discretion than to rules. Third, if the policymaker's rate of time preference rises during emergencies, the theory predicts higher monetary uncertainty and higher monetary variability at such times. This seems to be consistent with numerous wartime episodes.

The model of this chapter identifies some of the determinants of credibility. The chapter also suggests that the so-called "credibility hypothesis" and the rational expectations paradigm should be viewed as complements rather than as substitutes.[32]

Appendix

1. *Derivation of the optimal predictor in equation (10) of the text*

Define the dummy stochastic variable $\varepsilon_t \equiv \psi_t/B$. Substituting this relation into equation (8)

$$y_t = B(p_t + \varepsilon_t) \equiv By_t^1$$

---

32. Fellner and Haberler, who introduced the term *credibility hypothesis*, seem to imply that they are at least partial substitutes (see Fellner, 1976, p. 170; 1979). Haberler (1980, p. 280), wrote: "Thus the credibility approach does not accept the assumption made by the rational expectations school that government actions fall nearly into two extreme categories—systematic, wholly predictable policies and unsystematic, entirely unpredictable shocks. Especially after a long period of inflation, which has shaken the public's confidence that the government will carry out its anti-inflationary

Since the public knows the parameter B, an observation on y_t is equivalent to an observation on y_t^1. Hence the expected value of p_t conditioned on past values of y is equal to this expected value conditioned on past values of y^1. We turn now to the calculation of this expected value.

Since p_t and y_t^1 are normally distributed, the expected value of p_t conditioned on y_{t-i}^1, $i \geq 1$, is a linear function with fixed coefficients of the observations on y_{t-i}^1, $i \geq 1$. That is

$$E[p_t | y_{t-1}^1, y_{t-2}^1, \ldots] = \sum_{i=1}^{\infty} a_i y_{t-i}^1 \tag{A1}$$

Since this conditional expected value is also the point estimate of p_t, which minimizes the mean square error around this estimate, it follows that $\{a_i\}_{i=1}^{\infty}$ are to be chosen to minimize

$$Q \equiv E\left[p_t - \sum_{i=1}^{\infty} a_i y_{t-i}^1\right]^2. \tag{A2}$$

Using the relation between y and y^1 in (A2), using the fact that ε and v are mutually independent, and passing the expectation operator through, Q can be rewritten

$$Q = [1 + (\rho - a_1)^2 + (\rho^2 - \rho a_1 - a_2)^2 + \cdots + (\rho^i - \rho^{i-1} a_1 - \cdots - a_i)^2 + \cdots]\sigma_v^2$$
$$+ \sum_{i=1}^{\infty} a_i^2 \sigma_\varepsilon^2 \quad \text{where } \sigma_\varepsilon^2 = \frac{\sigma_\psi^2}{B^2}. \tag{A3}$$

The necessary first-order conditions for an extremum of Q are

$$\frac{\partial Q}{\partial a_i} = -2[(\rho^i - \rho^{i-1} a_1 - \cdots - a_i) + \rho(\rho^{i+1} - \rho^i a_1 - \cdots - a_{i+1})$$
$$+ \rho^2(\rho^{i+2} - \rho^{i+1} a_1 - \cdots - a_{i+2}) + \cdots] + 2\sigma_\varepsilon^2 a_i = 0, \quad i \geq 1 \tag{A4}$$

Leading (A4) by one period, multiplying by ρ, and subtracting (A4) from the resulting expression

$$(\rho^i - \rho^{i-1} a_1 - \cdots - a_i)\sigma_v^2 + (\rho a_{i+1} - a_i)\sigma_\varepsilon^2 = 0, \quad i \geq 1. \tag{A5}$$

Multiplying (A5) by ρ, substracting the resulting expression from (A5) led by one

---

policy, a sustained and deliberate effort must be made to restore credibility." The analysis presented in our chapter suggests that it is possible to accept Haberler's second statement without having to accept the first one. High credibility will quicken and ease the process of disinflation even in a rational expectations model in which policy actions can be decomposed—as is the case here—into predictable and unpredictable components. Fellner softened his position in Fellner (1980).

period, using the relation $\sigma_\varepsilon^2 = \sigma_\psi^2/B^2$, and rearranging,

$$a_{i+2} - \left(\frac{1+r}{\rho} + \rho\right)a_{i+1} + a_i = 0, \, i \geq 1 \quad \text{where } r = \frac{\sigma_v^2}{\sigma_\varepsilon^2} = \frac{B^2\sigma_v^2}{\sigma_\psi^2} \tag{A6}$$

This is a second-order homogeneous difference equation whose general solution is

$$a_i = C\lambda^i \tag{A7}$$

where C is a constant to be determined by initial conditions and λ is the root of the quadratic

$$u^2 - \left(\frac{1+r}{\rho} + \rho\right)u + 1 = 0. \tag{A8}$$

The roots of this equation are given by

$$u_{1,2} = \frac{1}{2}\left(\frac{1+r}{\rho} + \rho\right) \pm \sqrt{\left[\frac{1}{2}\left(\frac{1+r}{\rho} + \rho\right)\right]^2 - 1} \tag{A9}$$

The positive root in (A9) is larger than one and the negative root is bounded between zero and one. Thus a_i does not diverge only if the smaller root is substituted for λ in (A7). Since a_i has to yield a minimum for Q it cannot diverge. Hence

$$\lambda = \frac{1}{2}\left(\frac{1+r}{\rho} + \rho\right) - \sqrt{\left[\frac{1}{2}\left(\frac{1+r}{\rho} + \rho\right)\right]^2 - 1}.$$

For $i = 1$, (A5) implies

$$(\rho - a_1)\sigma_v^2 + (\rho a_2 - a_1)\sigma_\varepsilon^2 = 0 \tag{A10}$$

Using (A7) to express a_1 and a_2 in terms of C and λ, substituting into (A10), and rearranging

$$C = \frac{\rho\sigma_v^2}{\lambda[\sigma_v^2 + \sigma_\varepsilon^2 - \rho\lambda\sigma_\varepsilon^2]} = \frac{\rho}{\lambda}\frac{r}{1+r-\rho\lambda} \tag{A11}$$

Since λ is a root of the quadratic in (A8) it satisfies

$$\lambda^2 - \left(\frac{1}{\rho} + \rho + \frac{r}{\rho}\right)\lambda + 1 = 0$$

which implies

$$r = \rho\lambda + \frac{\rho}{\lambda} - (1 + \rho^2) \tag{A12}$$

Substituting (A12) into (A11) and rearranging

$$C = \frac{\rho - \lambda}{\lambda}. \tag{A13}$$

Substituting (A13) into (A7), substituting the resulting expression into (A1), and rearranging we obtain

$$E[p_t | y^1_{t-1}, y^1_{t-2}, \ldots] = (\rho - \lambda) \sum_{j=0}^{\infty} \lambda^j y^1_{t-1-j}$$

$$= \frac{\rho - \lambda}{B} \sum_{j=0}^{\infty} \lambda^j (B p_{t-1-j} + \psi_{t-1-j}) = \frac{\rho - \lambda}{B} \sum_{j=0}^{\infty} \lambda^j y_{t-1-j}$$

The optimal predictor in equation (10) of the text follows by recalling that past values of y^1 and y carry the same information. Hence

$$E[p_t | y^1_{t-1}, y^1_{t-2}, \ldots] = E[p_t | y_{t-1}, y_{t-2}, \ldots] = \frac{\rho - \lambda}{B} \sum_{j=0}^{\infty} \lambda^j y_{t-1-j}$$

which is equation (10a) in the text.

Note that for $\rho = 1$ and $B = 1$, p_i becomes a random walk and this predictor reduces to the Muth (1960) optimal predictor.

2. Proof of Proposition 3

Substituting (7) and (9) into the expression for $V(e)$ that precedes proposition 3 and rearranging

$$V(e) = \sigma_\psi^2 + B^2 E[p_i - E[p_i | y_{i-1}, y_{i-2}, \ldots]]^2. \tag{A14}$$

Since $E[p_i | y_{i-1}, y_{i-2}, \ldots]$ is a minimum mean square error predictor

$$E[p_i - E[p_i | y_{i-1}, y_{i-2}, \ldots]] = \min_{\{a_j\}} E[p_i - \sum_{j=1}^{\infty} a_j y_{i-j}]^2. \tag{A15}$$

Substituting (A15) into (A14), passing B to the right of the min operator, and redefining variables

$$V(e) = \sigma_\psi^2 + \min_{\{a_j\}} E[q_i - \sum_{j=1}^{\infty} a_j (q_{i-j} + \psi_{i-j})]^2$$

$$q_j \equiv B p_j \quad \text{for all } j. \tag{A16}$$

The definition of q_j in (A16) in conjunction with (5) implies

$$Eq = 0, \quad \sigma_q^2 = B^2 \sigma_p^2, \quad Eq_i q_{i-j} = \rho^j \sigma_q^2 \qquad (A17)$$

Expanding the square term on the right hand side of (A16), using (A17) and the fact that q_j and ψ_j are mutually and serially uncorrelated, equation (A16) can be rewritten

$$V(e) = \sigma_\psi^2 + \min_{\{a_j\}} \left[\sum_{j=1}^{\infty} a_j^2 \sigma_\psi^2 + Q[\{a_j\}] \sigma_q^2 \right] \qquad (A18)$$

where

$$Q[\{a_j\}] \equiv \left(1 - 2 \sum_{j=1}^{\infty} a_j \rho^j + \sum_{j=1}^{\infty} \sum_{t=1}^{\infty} a_j a_t \rho^{|j-t|} \right). \qquad (A19)$$

The difference between the minimization problems in (A15) and (A16) is only that in the later case the constant B has been absorbed into the minimization problem. Hence the minimizing values of $\{a_i\}$ are identical for the two problems and are (see (10)) given by

$$a_j = (\rho - \lambda)\lambda^{j-1}. \qquad (A20)$$

Substituting (A20) into (A19) and rearranging

$$Q[\{a_j\}] = \frac{1 - \rho^2}{1 - \lambda^2} > 0. \qquad (A21)$$

By the envelope theorem, the sign of the total effect of a change in σ_q^2 on $\dot{V}(e)$ in (A18) is the same as that of the direct effect of σ_q^2, which is given by $Q[\cdot]$ in (A21). Since $Q[\cdot]$ is positive, this implies that the total effect of a decrease in σ_q^2 is to decrease $V(e)$. But we saw in footnote 16 that an increase in β reduces B, which reduces σ_q^2 through (A17). It follows that an increase in β reduces $V(e)$.

3. Derivation of Equation (20)

Collecting all the terms involving $B_0 A$ in (19) it is easily seen that their sum is zero. Substituting (4) and (6) into the remaining expression and taking its unconditional expected value we obtain

$$EJ(\cdot) = \sum_{i=0}^{\infty} \beta^i \left\{ BE\left[p_i - (\rho - \lambda) \sum_{j=0}^{\infty} \lambda^j p_{i-j-1} \right](A + p_i) - \frac{1}{2} E(B_0 A + B p_i)^2 \right\} \qquad (A22)$$

In deriving (A22) use has been made of the fact that ψ has a zero expected value and is statistically independent of p. Passing the expectation operator in (A22) through and recalling that $Ep_i = 0$, $Ep_i^2 = \sigma_p^2$, and $Ep_i p_{i-j} = \rho^j \sigma_p^2$, we obtain

$$EJ(\cdot) = \sum_{i=0}^{\infty} \beta^i \left\{ B\sigma_p^2 [1 - (\rho - \lambda)(\rho + \lambda \rho^2 + \lambda^2 \rho^3 + \ldots)] - \frac{1}{2}(B_0^2 A^2 + B^2 \sigma_p^2) \right\} \quad \text{(A23)}$$

Summing the infinite sums involving β and $\lambda \rho$ and using (16), (A23) can be rewritten

$$\frac{1}{1-\beta} \left[\frac{1-\beta\rho^2}{1-\beta\rho\lambda} \frac{1-\rho^2}{1-\lambda\rho} \sigma_p^2 - \left(\frac{1-\beta\rho^2}{1-\beta\rho\lambda}\right)^2 \frac{\sigma_p^2}{2} - \frac{1}{2}\left(\frac{1-\beta\rho}{1-\beta\lambda}\right)^2 A^2 \right] \quad \text{(A24)}$$

Equation (20) in the text follows by noting that $\sigma_p^2 = \sigma_v^2/(1-\rho^2)$, $K^2 = A^2/\sigma_p^2$, and by rearranging.

4. Proof of Proposition 4

It is convenient to start by rewriting equation (23) of the text as

$$\frac{1-\beta}{\sigma_v^2} F(\lambda) = A_D(\lambda) - B_D(\lambda) - C_D(\lambda)K^2 \equiv F_1(\lambda) \quad \text{(A25)}$$

where

$$A_D(\lambda) \equiv \frac{\rho(1-\beta\rho^2)(1+\beta-2\beta\rho\lambda)}{(1-\beta\rho\lambda)^2(1-\rho\lambda)^2} \quad \text{(A26a)}$$

$$B_D(\lambda) \equiv \frac{\beta\rho(1-\beta\rho^2)^2}{(1-\rho^2)(1-\beta\rho\lambda)^3} \quad \text{(A26b)}$$

$$C_D(\lambda) \equiv \frac{\beta(1-\beta\rho)^2}{(1-\rho^2)(1-\beta\lambda)^3} \quad \text{(A26c)}$$

Lemma: $A_D(\lambda)$, $B_D(\lambda)$, and $C_D(\lambda)$ are all monotonically increasing functions of λ.

Proof of Lemma: The lemma follows by inspection of (A26a) and (A26c) and by noting that the partial derivative of $A_D(\lambda)$ with respect to λ is

$$\frac{\partial A_D(\lambda)}{\partial \lambda} = \frac{2\rho^2(1-\beta\rho^2)\{(1-\rho\beta\lambda)^2 + [\beta(1-\rho\lambda)]^2 + \beta(1-\rho\lambda)(1-\rho\beta\lambda)\}}{[(1-\beta\rho\lambda)(1-\rho\lambda)]^3} \quad \text{(A27)}$$

Since β, ρ, and λ are all between zero and one, the partial derivative in (A27) is positive. This completes the proof of the lemma.

If $F(\lambda)$ is negative for all $0 \leq \lambda \leq \rho$, $G(\lambda)$ is decreasing in λ in all the relevant range and $\lambda^* = 0$. Since $\beta < 1$ and $\sigma_v^2 > 0$, $F(\lambda)$ and $F_1(\lambda)$ have the same sign. It follows that a sufficient condition for $\lambda^* = 0$ is $F_1(\lambda) < 0$ for all $0 \leq \lambda \leq \rho$. The lemma implies

$$F_1(\lambda) \leq A_D(\rho) - B_D(0) - C_D(0)K^2 \equiv H_1, \quad 0 \leq \lambda \leq \rho \tag{A28}$$

Hence a sufficient condition for $\lambda^* = 0$ is $H_1 < 0$ which, using (A28), is equivalent to

$$K^2 > c_1. \tag{A29}$$

Similarly a sufficient condition for $\lambda^* = \rho$ is $F_1(\lambda) > 0$ for all $0 \leq \lambda \leq \rho$. The lemma implies

$$F_1(\lambda) \geq A_D(0) - B_D(\rho) - C_D(\rho)K^2 \equiv H_2, \quad 0 \leq \lambda \leq \rho \tag{A30}$$

Hence a sufficient condition for $\lambda^* = \rho$ is $H_2 > 0$ which, using (A30), is equivalent to

$$K^2 < c_2. \tag{A31}$$

We prove that $c_1 \geq c_2$ by contradiction. Suppose $c_1 < c_2$; then there exists a value of K^2 such that $c_1 < K^2 < c_2$. Since $K^2 > c_1$, (A21) implies that $\lambda^* = 0$. But since $K^2 < c_2$, (A31) implies that $\lambda^* = \rho$, which is a contradiction. This establishes parts (i) through (iii) of the proposition.

A sufficient condition for an internal value of λ^* is

$$F_1(0) > 0 \quad \text{and} \quad F_1(\rho) < 0. \tag{A32}$$

Using (A25) in (A32) and rearranging, this sufficient condition can be reformulated as

$$c_L < K^2 < c_H \tag{A33}$$

We prove that when $c_H > c_L$, both of those numbers must be between c_1 and c_2 by contradiction. Suppose (A33) is satisfied but that c_L is lower than c_2. Then there exists a K^2 that satisfies both (A31) and (A33), implying that λ^* is internal and also equal to ρ, which is a contradiction. Alternatively, suppose (A33) holds and $c_H > c_1$. Then there exists a K^2 that satisfies both (A29) and (A33), implying that λ^* is both internal and equal to 0, which is a contradiction. It follows that when $c_L < c_H$, both c_L and c_H must be bounded between c_2 and c_1. This establishes part (iv) of the proposition.

References

Backus, D. and Driffill, J. (1985). "Inflation and Reputation." *American Economic Review*, 75, pp. 530–38.

Barro, R. J. and Gordon, D. B. (1983a). "Rules, Discretion and Reputation in a Model of Monetary Policy." *Journal of Monetary Economics*, 12, pp. 101–22.

Barro, R. J. and Gordon, D. B. (1983b). "A Positive Theory of Monetary Policy in a Natural Rate Model." *Journal of Political Economy*, 91, pp. 589–610.

Bomhoff, E. J. (1982). "Predicting the Price Level in a World that Changes All the Time." *Carnegie-Rochester Conference Series on Public Policy*, 17 (Autumn), pp 7–57.

Brunner, K., Cukierman, A., and Meltzer, A. H. (1983). "Money and Economic Activity, Inventories and Business Cycles." *Journal of Monetary Economics*, 11 (May), pp. 281–320.

Brunner, K. and Meltzer, A. H. (1964). *The Federal Reserve Attachment to Free Reserves*. Washington, D.C.: House Committee on Banking and Currency.

Bull, C. (1982). "Rational Expectations, Monetary Data and Central Bank Watching." *Giornale degli Economisti e Annali di Economia*, 41 (Jan.-Feb.), pp. 31–40.

Burns, A. F. (1979). *The Anguish of Central Banking*. Belgrade, Yugoslavia: Per Jacobsson Foundation.

Canzoneri, M. B. (1985). "Monetary Policy Games and the Role of Private Information." *American Economic Review*, 75 (Dec.), pp. 1056–70.

Cukierman, A. (1984). *Inflation, Stagflation, Relative Prices and Imperfect Information*. Cambridge: Cambridge University Press.

Cukierman, A. and Meltzer, A. H. (1986). "The Credibility of Monetary Announcements." In M. J. M. Neumann, ed., *Monetary Policy and Uncertainty*. Baden Baden: Nomos Publishing.

Fellner, W. (1976). *Towards a Reconstruction of Macroeconomics: Problems of Theory and Policy*. Washington, D.C.: American Enterprise Institute.

Fellner, W. (1979). "The Credibility Effect and Rational Expectations: Implications of the Gramlich Study." *Brookings Papers on Economic Activity*, No. 1, pp. 167–78.

Fellner, W. (1980). "The Valid Core of Rationality Hypotheses In the Theory of Expectations." *Journal of Money Credit and Banking*, 12 (Nov.), pp. 763–87.

Fischer, S. (1977). "Long Term Contracts, Rational Expectations, and the Optimal Money Supply Rule." *Journal of Political Economy*, 85 (Apr.), pp. 191–206.

Fischer, S. (1984). Contracts, Credibility and Disinflation. Working Paper No. 1339, National Bureau of Economic Research, April.

Fischer, S. and Huizinga, J. (1982). "Inflation, Unemployment, and Public Opinion Polls." *Journal of Money, Credit and Banking*, 14 (Feb.), pp. 1–19.

Frey, B. S. and Schneider, F. (1978). "An Empirical Study of Politico-Economic Interactions in the U.S." *Review of Economics and Statistics*, 60 (May), pp. 174–83.

Friedman, M. (1960). *A Program for Monetary Stability*. New York: Fordham University Press.

Goodfriend, M. (1986). "Monetary Mystique: Secrecy and Central Banking." *Journal of Monetary Economics*, 17 (Jan.), pp. 63–92.

Goodhart, C. A. E. and Bhansali, R. J. (1970). "Political Economy." *Political Studies*, 18, pp. 43–106.

Green, E. J. and Porter, R. H. (1984). "Noncooperative Collusion under Imperfect Price Information." *Econometrica*, 52 (Jan.), pp. 87–100.

Haberler, G. (1980). "Notes on Rational and Irrational Expectations." Washington, D.C.: American Enterprise Institute, Reprint no. 11, March 1980. Reprinted from Emil

Kung, ed., *Wandlungen in Wirtschaft und Gesellschaft: Die Wirtschafts und die Socialwissenschaften vor neuen Aufgaben*. Turbingen: J. C. B. Mohr.

Hardouvelis, G. (1984). "Market Perceptions of Federal Reserve Policy and the Weekly Monetary Announcements." *Journal of Monetary Economics, 14* (Sept.), pp. 225–40.

Hetzel, R. L. (1984). The Formulation of Monetary Policy in a Democracy. Unpublished manuscript, Federal Reserve Bank of Richmond, January.

Jaffe, D. M. and Kleiman, E. (1977). "The Welfare Implications of Uneven Inflation." In E. Lundberg, ed., *Inflation Theory and Anti-Inflation Policy*. London: Macmillan.

Kane, E. J. (1982). "External Pressure and the Operation of the Fed." In *Political Economy of International and Domestic Monetary Relations*, R. E. Lombra and W. E. Witte, eds. Ames: Iowa University Press. pp. 211–32.

Kydland, F. E. and Prescott, E. C. (1977). "Rules Rather than Discretion: The Inconsistency of Optimal Plans." *Journal of Political Economy, 85* (June), pp. 473–92.

Logue, D. E. and Willet, T. D. (1976). "A Note on the Relation Between the Rate and the Variability of Inflation." *Economica, 43*, pp. 151–58.

Lombra, R. and Moran, M. (1980). "Policy Advice and Policymaking at the Federal Reserve." In K. Brunner and A. H. Meltzer, eds., *Monetary Institutions and the Policy Process*, Carnegie-Rochester Conference Series on Public Policy, *15*, pp. 9–68.

Lucas, R. E. Jr. (1973). "Some International Evidence on Output Inflation Tradeoffs." *American Economic Review, 63* (June), pp. 326–35.

Meyer, L. H. and Webster, C. Jr. (1982). "Monetary Policy and Rational Expectations: A Comparison of Least Squares and Bayesian Learning." *Carnegie-Rochester Conference on Public Policy, 17* (Autumn), pp. 67–99.

Muth, J. F. (1960). "Optimal Properties of Exponentially Weighted Forecasts." *Journal of the American Statistical Association, 35* (June), pp. 299–306.

Okun, A. M. (1971). "The Mirage of Steady Anticipated Inflation." *Brookings Papers on Economic Activity, 2*, pp. 485–98.

Sargent, T. J. (1979). *Macroeconomic Theory*. New York: Academic Press.

Taylor, J. B. (1980a). "Aggregate Dynamics and Staggered Contracts." *Journal of Political Economy, 88*, pp. 1–23.

Taylor, J. B. (1980b). "Recent Developments in the Theory of Stabilization Policy." In Lawrence Meyer, ed., *Stabilization Policies; Lessons from the Seventies and Implications for the Eighties*. Center for the Study of American Business. Conference Vol. 4, Federal Reserve Bank of St. Louis, April, pp. 1–40.

Taylor, J. B. (1982). "Establishing Credibility: A Rational Expectations Viewpoint." *American Economic Review, 72* (May), pp. 81–5.

Taylor, J. B. (1983). "Comment on Rules, Discretion and Reputation in a Model of Monetary Policy." *Journal of Monetary Economics, 12* (July), pp. 123–25.

Weintraub, R. (1978). "Congressional Supervision of Monetary Policy." *Journal of Monetary Economics, 4* (Apr.), pp. 341–62.

Woolley, J. T. (1984). *The Federal Reserve and the Politics of Monetary Policy*. New York: Cambridge University Press.

9

The Credibility of Monetary Announcements

ALEX CUKIERMAN AND ALLAN H. MELTZER

I. Introduction

In the mid-1970s, several countries began to announce targets for monetary growth. U.S. Congressional Resolution 133 required the Federal Reserve to provide the public and Congress with more precise information about the monetary actions it contemplated than it had previously provided. Other countries adopted similar procedures.[1] The results have been mixed. Some countries have maintained money growth close to the announced target. Others have not. The announcements have not proven to be reliable guides to actual money growth. Should they be abandoned?

This chapter investigates the relationship between the credibility of monetary announcements and the structure of the policymaker's objectives when the policymaker must make announcements but is not bound by any precise rule and is not required to make precise announcements.[2] The chapter shows that when the policymaker has an information advantage about his own shifting objectives, and policy is discretionary, announcements will neither be completely credible nor completely incredible. The importance given by the public to forecasts of future monetary trends does not depend on the absolute degree of precision of announcements. It depends instead on the relative degree of noise in the announcements and in the control of money supply. With imprecise monetary control, relatively more precise announcements will command high credibility even if they are absolutely imprecise.

Two measures for the credibility of announcements are proposed. The first—*average credibility*—is inversely related to the distance between the

---

1. At the time of writing, announcements have been made in Germany, Japan, United Kingdom, France, Canada, Switzerland, Italy, Netherlands, and the United States.
2. In practice, the Federal Reserve announces a range for money growth.

current announcement and the public's beliefs. The second—*marginal credibility*—measures the extent to which a one-unit change in announced targets affects expectations. Both measures are negatively related to the relative noise in announcements and in monetary control. Average credibility is also reduced when the objectives of the monetary authority undergo large changes.[3] Average credibility has the attribute of a capital good that is built up and depleted gradually. The more limited is the control of money by the policymaker and the noisier his announcements, the longer it takes to build up and to deplete the "stock" of credibility.

Since announcements have been instituted to provide the public and Congress with more information about future monetary policy, it is important to determine whether they reduce the public's uncertainty. The chapter gives an affirmative answer to this question by showing that preannouncement of monetary targets never increases the level of monetary uncertainty and usually decreases it. In this sense the requirement to disclose targets is probably useful to the public. Other results are:

1. A Friedman (1960)-type constant rate of money growth rule decreases the level of monetary uncertainty even below the level achieved under discretionary policy with preannouncement of targets.
2. Preannouncement of targets reduces the level of monetary variability[4] but not to the level that could be attained with a constant planned rate of money growth.
3. When policymakers must make completely credible announcements, they are induced to plan a constant rate of money growth even if he has the discretion to do otherwise. The rate they choose is zero. (There is no growth of output).
4. The larger the degree of time preference of policymakers, the larger the monetary uncertainty experienced by the public.

Section II presents the model of central bank behavior and the formation of expectations by the public. The decision rule of the policymaker and the rationality of expectations are discussed in Section III. Section IV presents measures of credibility of announcements and relates them to some underlying parameters of the environment. A comparison of monetary variability and monetary uncertainty in the presence and in the absence of announcements appears in Section V. Concluding remarks follow.

---

3. Credibility here refers to the degree to which announcements are believed. It is imperfect because of imprecise announcements rather than because of dynamic inconsistency of the type discussed by Kydland and Prescott (1977) and Barro and Gordon (1983b). Further differences between those two concepts of credibility and their underlying conceptual framework are discussed in Cukierman (1986).

4. Variability and uncertainty are not necessarily identical. See, for example, Cukierman (1984), Chapter 4, Section 4.

II. The Model

The model in this section is a generalization of the model presented in Cukierman and Meltzer (1986) (CM in what follows and represented as Chapter 8). This model features a monetary authority whose relative preference for different objectives, such as economic stimulation via monetary surprises or inflation prevention, shifts in a stochastic manner over time. The shift changes the emphasis given to particular objectives during choice of money growth.

Despite the statutory independence of many central banks, their policies are partially responsive to the conflicting desires of other governmental institutions. Students of the Fed suggest that the formulation of monetary policy in the United States is not divorced from the general democratic process (Hetzel, 1984; Kane, 1980, 1982; Weintraub, 1978; Wooley, 1984). The Fed is partly responsive to the desires of the president (Beck, 1982) and to pressures from Congress—the body to which it is formally responsible.[5] However, because of its statutory independence, the Fed has some authority to decide whose wishes to accommodate first and how much to accede to pressure.

The model has two crucial features. First, the Federal Reserve is better informed than the public about its objectives at a given time. Since the Fed's objectives are private information, the public does not learn about changes in objectives directly. Second, monetary control is imperfect, so the public cannot immediately infer the Fed's objectives by observing the rate of money growth. Persistent changes in money growth, resulting from changes in objectives, are mixed with control errors. By observing money growth the public gradually and rationally learns about the emphasis given to different objectives of monetary policy.

The model is rational in the sense that the actual behavior that emerges is the same as the behavior on which the public relies to form expectations about future money growth. The policymaker knows how the public forecasts money growth and inflation, so he can calculate, up to a random shock, the effect of a given choice of money growth on surprise creation. The policymaker chooses the rate of money growth by comparing the benefits (to the policymaker) from surprise creation against the costs of higher inflation.

Each period the policymaker plans to achieve a certain rate of money growth, m_i^p. Actual money growth, m_i, may differ from the planned rate because control is imperfect. Specifically

$$m_i = m_i^p + \psi_i \tag{1}$$

The variance σ_ψ^2 of the noise term in (1) reflects the extent to which the operating procedures and the institutional environment prevent perfect control of money growth.

---

5. Part of those desires are motivated by political-distributional considerations. A summary view of the political approach to central bank behavior appears in Section II of Cukierman (1986).

The policymaker's decision-making strategy is

$$\max_{\{m_i^p,\, i=0,1,\ldots\}} E_{G0} \sum_{i=0}^{\infty} \beta^i \left[e_i x_i - \frac{(m_i^p)^2}{2} \right] \quad (2)$$

$$e_i \equiv m_i - E[m_i | I_i] \quad (3)$$

$$x_i = A + p_i \qquad A > 0 \quad (4)$$

$$p_i = \rho p_{i=1} + v_i, \qquad 0 < \rho < 1 \quad (5)$$

where v is a serially uncorrelated normal variate with zero mean and variance σ_v^2 and is distributed independently of the control error ε. Here e_i is the unanticipated rate of money growth in period i, I_i the information available to the public at the beginning of period i, and $E[m_i|I_i]$ the public's forecast of m_i given the information set I_i. β is the policymaker's subjective discount factor, and E_{G0} is a conditional expected-value operator that is conditioned on the information available to government in period 0 including a direct observation on x_0. Ceteris paribus the policymaker prefers lower to higher inflation.

x_i is a randon shift parameter that determines the shifts in the policymaker's objectives between economic stimulation achieved through surprise creation and inflation. The higher the value of x_i the more willing is the policymaker to trade higher inflation for more stimulation. Equations (4) and (5) specify the stochastic behavior of the shift parameter x_i and indicate that the policymaker's objectives exhibit a certain degree of persistence. The degree of persistence depends on the size of A and ρ. x_i is normally positive,[6] reflecting the view that unanticipated monetary growth stimulates employment and output and that the policymaker prefers, ceteris paribus, more rather than less stimulation.

House Concurrent Resolution 133, and later the Humphrey-Hawkins Act, requires the Federal Reserve to announce planned rates of growth for principal monetary aggregates. The purpose of this legislation is to provide the public and Congress with more precise information about the particular monetary actions contemplated by the monetary authority. This information has not proven to be a reliable forecast. Even a cursory look at the "Record of Policy Actions of the Federal Open Market Committee"[7] shows that actual rates of monetary growth have deviated substantially from the preannounced rates.

The main novelty of this Chapter is that it extends the model of CM to the case in which the policymaker must make announcements of future targets but is not required to make completely accurate announcements. We assume that at

---

6. The distributional assumptions on v imply that $x \sim N[A, \sigma_v^2/(1-\rho^2)]$ so that if $A\sqrt{1-\rho^2}/\sigma_v$ is sufficiently large, the probability that x will have a negative realization can be driven as close to zero as desired. For example, when this ratio is 3 the probability is 0.0027. To avoid negative values, A has to be sufficiently large compared with $\sigma_v/\sqrt{1-\rho^2}$.

7. The "Record" appears in the Federal Reserve Bulletin within two months after the meeting of the Federal Open Market Committee. Brunner and Meltzer (1983) have summarized the record.

the beginning of each period the policymaker makes a noisy announcement,

$$m_i^a = m_i^p + \gamma_i \qquad (6)$$

where γ_i is a serially uncorrelated normal variate with zero mean and variance σ_γ^2 and is distributed independently of the monetary control error.[8]

Preannouncement gives the public an additional source of information about future monetary growth. The optimal predictor incorporates the information that the public gains from the announcement and, as in CM, the information obtained from the history of money growth. A rational policymaker must use the new information structure faced by the public when planning future money growth. To obtain a rational-expectations solution for the model in the presence of announcements, we extend our earlier model to include announcements. As before, we postulate that the policymaker's strategy is given by

$$m_i^p = B_0 A + B p_i \qquad (7)$$

where B_0 and B are constants to be determined by rational expectations.

The public knows the structure of the policymaker's decisions regarding m_i^p, given by (7), and the relationship between announced and planned monetary growth in (6). The public observes only actual and announced money growth. At the beginning of period i the public knows the history of rates of money growth up to and including period $i - 1$ and the history of announcements up to and including the announcement, m_i^a, made at the beginning of period i. I_i denotes this information set.

Substituting (7) into (1) and (6), we have

$$m_i = B_0 A + B p_i + \psi_i \equiv B_0 A + y_i \qquad (8a)$$
$$m_i^a = B_0 A + B p_i + \delta_i \equiv B_0 A + z_i \qquad (8b)$$

Since B_0 and A are known parameters, observations on m_i and m_i^a amount to observations on y_i and z_i, respectively. Note that it pays the public to use announcements to improve its forecast of the stochastic persistent component of the policymaker's objectives.

It follows, using (8a), that the public's rational expectation of m_i is

$$E[m_i|I_i] = B_0 A + B E[p_i|I_i] = B_0 A + B E[p_i|z_i, z_{i-1}, \ldots, y_{i-1}, y_{i-2}, \ldots] \qquad (9)$$

---

8. Our formulation reflects the procedures in the United States where the Federal Reserve decides on the content of the announcement and chooses to announce a range rather than a point target.

Part 1 of the appendix shows that

$$E[p_i|I_i] = \frac{(\rho - \delta)(\sigma_\psi^2 + \sigma_\gamma^2)}{(\rho\sigma_\psi^2 + \delta\sigma_\gamma^2)B} \left\{ \sum_{j=1}^{\infty} \delta^j[\theta y_{i-j} + (1-\theta)z_{i-j}] + (1-\theta)z_i \right\} \tag{10a}$$

$$\delta \equiv \frac{1}{2}\left(\frac{1+r}{\rho} + \rho\right) - \sqrt{\frac{1}{4}\left(\frac{1+r}{\rho} + \rho\right)^2 - 1} \tag{10b}$$

$$r \equiv B^2 \frac{\sigma_v^2}{\sigma_\psi^2}\left(1 + \frac{\sigma_\psi^2}{\sigma_\gamma^2}\right) = \left(\frac{\sigma_v^2}{\sigma_\psi^2} + \frac{\sigma_v^2}{\sigma_\gamma^2}\right)B^2 \tag{10c}$$

$$\theta \equiv \frac{\sigma_\gamma^2}{\sigma_\psi^2 + \sigma_\gamma^2} \tag{10d}$$

Substituting (10a) into (9) and using (8)

$$E[m_i|I_i] = \frac{(\rho-\delta)(1-\theta)}{\delta + (\rho-\delta)(1-\theta)} m_i^a + \frac{\delta}{\delta + (\rho-\delta)(1-\theta)}$$

$$\times \sum_{j=0}^{\infty} \delta^j \{(1-\rho)\bar{m}^p + (\rho-\delta)[\theta m_{i-1-j} + (1-\theta)m_{i-1-j}^a]\} \tag{11}$$

where

$$\bar{m}^p \equiv B_0 A \tag{12}$$

is the mean rate of money growth.

Since δ has the same form in ρ and r as does λ in equation (10b) of CM, we can apply our previous result. (The solution for r differs in the two cases, since implicitly σ_γ^2 is infinity in CM). It follows that $0 \leq \delta \leq 1$ and $\rho - \delta \geq 0$. The optimal predictor is a weighted average of the current announcement and the history of announcements and monetary growth (including $\bar{m}^p$)—the two terms in (11)—with weights that sum to unity. The weight placed on each term depends on σ_γ^2. Noisy announcements—large σ_γ^2—reduce the usefulness of announcements, so the public pays less attention to announcements. In the limit as $\sigma_\gamma^2 \to \infty$, $\theta \to 1$; there is no information in the announcements and they are ignored. (The optimal predictor in (11) reduces to equation (11a) of CM. At the other extreme, $\sigma_\gamma^2 \to 0$,[9] announcements are completely accurate statements of planned money growth, and the optimal predictor, $E[m_i|I_i]$, equals m^a. The current announcement is fully credible. This remains true even if σ_ψ^2 is relatively large so that monetary control is relatively poor.

Between the extreme values of σ_γ^2, σ_v^2 and σ_ψ^2 determine the weights given to

---

9. As $\sigma_\gamma^2 \to 0$, $\theta \to 0$, $r \to \infty$, and $\delta \to 0$.

announcements and to the history of monetary growth. In general, the noisier signal gets a smaller weight.[10]

III. The Policymaker's Decision Rule and Proof of the Rationality of Expectations

The policymaker knows that the public forms expectations according to (11), and he uses this information when choosing planned monetary growth. Substituting (11) into (3), substituting the resulting expression into (2), using (12), and rearranging, the maximization problem of the policymaker becomes

$$\max_{\{m_i^p, i=0,1,\ldots\}} E_{G0} \sum_{i=0}^{\infty} \beta^i \left\{ x_i \left[m_i^p + \psi_i - \frac{(1-\rho)\delta}{(1-\delta)(\rho-\delta)(1-\theta)+\delta} B_0 A \right.\right.$$
$$- \frac{(\rho-\delta)(1-\theta)}{(\rho-\delta)(1-\theta)+\delta}(m_i^p + \gamma_i)$$
$$\left.\left. - \frac{\delta(\rho-\delta)}{(\rho-\delta)(1-\theta)+\delta} \sum_{j=0}^{\infty} \delta^j (m_{i-1-j}^p + \psi_{i-1-j} + \gamma_{i-1-j}) \right] - \frac{(m_i^p)^2}{2} \right\} \quad (13)$$

The stochastic Euler equations necessary for an internal maximum of this problem are[11]

$$\left[1 - \frac{(\rho-\delta)(1-\theta)}{(\rho-\delta)(1-\theta)+\delta}\right] x_i - \frac{\delta(\rho-\delta)}{(\rho-\delta)(1-\theta)+\delta} E_{Gi}(\beta x_{i+1} + \beta^2 \delta x_{i+2} + \ldots)$$
$$- m_i^p = 0 \quad i = 0, 1, \ldots \quad (14)$$

CM show (equation (14)) that

$$E_{Gi} x_{i+j} = \rho^j x_i + (1-\rho^j) A, \quad j \geq 0 \quad (15)$$

Substituting (15) into (14) and using (4), this expression reduces after a considerable amount of algebra to (16).[12]

$$m_i^p = \frac{\delta(1-\rho\beta)(1+\sigma_\psi^2/\sigma_\gamma^2)}{(1-\delta\beta)[\delta + \rho(\sigma_\psi^2/\sigma_\gamma^2)]} A + \frac{(1-\beta\rho^2)\delta(1+\sigma_\psi^2/\sigma_\gamma^2)}{(1-\rho\beta\delta)[\delta + \rho(\sigma_\psi^2/\sigma_\gamma^2)]} p_i \quad (16)$$

Rationality of expectations implies that the coefficients of A and of p_i should be

---

10. The sign of the partial derivative of the coefficient of m_i^a with respect to σ_γ^2 is opposite to that of $\partial\delta/\partial\sigma_\gamma^2 + \delta\partial\theta/\partial\sigma_\gamma^2$. Since both δ and θ are increasing in σ_γ^2, the weight given to the current announcement is monotonically decreasing in σ_γ^2.

11. Since all the terms in (14) are finite, the transversatility condition is satisfied for any $\beta < 1$.

12. Details appear in part 2 of the appendix.

the same across equations (7) and (16), respectively, so

$$B_0 = \frac{(1-\beta\rho)(1+\sigma_\psi^2/\sigma_\gamma^2)\delta(B)}{[1-\beta\delta(B)][\delta(B)+\rho\sigma_v^2/\sigma_\gamma^2]}$$

$$B = \frac{(1-\beta\rho^2)(1+\sigma_\psi^2/\sigma_\gamma^2)\delta(B)}{[1-\beta\rho\delta(B)][\delta(B)+\rho\sigma_v^2/\sigma_\gamma^2]} \equiv G(B)$$

The dependence of δ on B (through equation (10)) is stressed by writing δ as a function of B. The second equation determines B uniquely as an implicit function of β, ρ, σ_v^2, σ_ψ^2, and σ_γ^2. Uniqueness is demonstrated by noting that the right-hand side of this equation is an increasing function of δ and that δ is a decreasing function of r (see footnote 15 of CM). Since, from (10c), r is increasing in B, $G(B)$ is therefore a monotonically decreasing function of B. Since $G(0) = 1$ this implies a unique positive solution for B.

When there is no information in the announcement ($\sigma_\gamma^2 \to \infty$), the decision strategy in (16) reduces to the decision strategy in the absence of announcements.[13] Hence both the optimal predictor and the policymaker's decision rule in the presence of announcements are generalizations of their respective counterparts in the economy with no announcements. Formally, therefore, the economy with no announcements can be viewed as a particular case (with $\sigma_\gamma^2 \to \infty$) of the economy with announcements.

From equation (8a) the unconditional mean and variance of actual money growth are given respectively by

$$Em_i = B_0 A \tag{17a}$$

$$V(m_i) = B^2 \frac{\sigma_v^2}{1-\rho^2} + \sigma_\psi^2 = B^2 \sigma_p^2 + \sigma_\psi^2 \tag{17b}$$

B_0 is positive,[14] so the average rate of monetary expansion and the average rate of inflation are positive. The average rates increase with the government's preference for economic stimulation relative to its dislike of monetary growth and inflation. Also for a given variance of the monetary control error, the variance of monetary growth is, as in CM, a decreasing function of β.[15] The more a policymaker cares about the present relative to the future, the more he increases the variability of monetary growth, independently of whether he makes or does not make announcements.

---

13. As $\sigma_\gamma^2 \to \infty$, δ tends to λ in CM, B_0 and B become identical to B_0 and B in CM, and the decision strategy in (16) reduces to that in equation (15) of CM. In addition, the optimal predictor in (11) reduces to equation (11a) of CM.

14. B_0 is positive for any $\beta < 1$, since $0 \geqslant \delta \geqslant 1$.

15. The proof is qualitatively similar to the proof of proposition 2b in CM.

IV. The Credibility of Announcements

To this point we have not given a precise definition of credibility. In this section we use the optimal predictor in (11) to define two measures of credibility, average and marginal credibility, and we analyze the determinants of credibility.

Average credibility measures the extent to which the public's expectations of future monetary growth differs from the current announcement. The smaller this difference, the larger the average credibility. When $m_i^a = E(m_i|I_i)$, average credibility is perfect.

$$\text{Average credibility} \equiv AC \equiv -|m_i^a - E[m_i|I_i]| \tag{18}$$

Substituting (11) into (18) and rearranging

$$AC = \frac{-\delta}{\delta + (\rho - \delta)(1 - \theta)}|m_i^a - m_i^*| \tag{19}$$

where

$$m_i^* \equiv \sum_{j=0}^{\infty} \delta^j \{(1 - \rho)\bar{m}^p + (\rho - \delta)[\theta m_{i-1-j} + (1 - \theta)m_{i-1-j}^a]\} \tag{20}$$

m_i^* summarizes all the information that individuals have about future money growth before they get the current announcement. Average credibility is low when the current announcement is far away from m_i^*. The distance between m_i^a and m_i^* rises with the difference between current and past announcements and with the difference between m_i^a and the mean rate of money growth $\bar{m}^p$. Average credibility is reduced both during periods with large changes in announcements and when announcements differ markedly from average experience. Further, for any given divergence between m_i^a and m_i^*, average credibility is lower the larger the coefficient

$$C \equiv \frac{\delta}{\delta + (\rho - \delta)(1 - \theta)}$$

in equation (19). This coefficient is, ceteris paribus, an increasing function of δ.[16] Analysis of C leads to the following proposition about average credibility.

Proposition 1: For a given divergence between m_i^a and m_i^*, average credibility is lower (a) the lower the precision of announcements (the larger the value of σ_y^2), (b) the lower equiproportionally the precision of both announcements and

---

16. The sign of its partial derivative with respect to δ is the same as the sign of $\rho(1 - \theta) > 0$.
17. The result in part (c) of the proposition has to be interpreted with care since it also reflects changes in the noisiness of monetary control and of announcements.

monetary control (the higher the value of σ_γ^2 and σ_ψ^2 for a given $\sigma_\gamma^2/\sigma_\psi^2$), and (c) the lower the variance of the innovation in the policymaker's objectives, σ_v^2.[17]

Proof: (a) Increases in σ_γ^2 increase δ and reduce $(1-\theta)$, so C increases with σ_γ^2. (b) An equiproportional increase in σ_ψ^2 and σ_γ^2 raises C by increasing δ for a fixed θ. (c) Follows directly from the fact that C increases with δ and δ decreases with σ_v^2. □

When monetary control is loose and announcements are highly noisy, a shift to new rates of monetary growth does not generate immediate credibility. In particular, a change in governmental objectives toward less inflation will be recognized only gradually even if government announces the new policy. As a result, the period of learning and the consequent lull in economic activity is lengthened; the cost of disinflation increases. A shift to more inflationary objectives also takes more time to be recognized when credibility is low, so the period of economic stimulation increases.

When monetary control is tight and announcements precise, the credibility of new objectives is quickly established. Note that when either monetary control is perfect ($\sigma_\psi^2 = 0$) or announcements are fully precise ($\sigma_\gamma^2 = 0$), $\delta = 0$ and average credibility is perfect. Hence perfect credibility can be established either way. Uncertainty faced by the public is larger, however, when the second method is used.

Average credibility focuses on the difference between the current announcement and belief. One may also be interested in the ability of the current announcement to affect expectations. A useful measure of this ability is marginal credibility.

$$\text{Marginal credibility} \equiv \text{MC} = \frac{\partial E[m_i | I_i]}{\partial m_i^a}$$

From equation (11)

$$\text{MC} = \frac{(\rho - \delta)(1 - \theta)}{\delta + (\rho - \delta)(1 - \theta)}$$

Considerations similar to those that led to proposition 1 show that marginal credibility falls with any increase in σ_γ^2, with any equiproportional increase in σ_γ^2 and σ_ψ^2, and with any decrease in σ_v^2. Marginal credibility is perfect when $\text{MC} = 1$ and is nonexistent when $\text{MC} = 0$. The first limiting case is attained for either $\sigma_\psi^2 \to 0$ or $\sigma_\gamma^2 \to 0$. The second is attained when both σ_ψ^2 and σ_γ^2 tend to infinity.

Before closing this section, we note that imperfect credibility arises here without any dynamic inconsistency of the type considered by Kydland and Prescott (1977, p. 475). The basis for imperfect credibility here is the policymaker's advantage over the public that is due to shifting objectives, noisy control, and noisy announcements. Policymakers know their stochastically changing objectives, but the public does not. The best the public can do is to form expectations, allowing for this noise, and use all the information available each period to infer current and future money growth. Announcements are not fully credible because they are noisy despite the fact that the policymaker is known to be following dynamically consistent discretionary policies.

V. Comparison of Uncertainty and Variability under Alternative Monetary Arrangements

In this section we compare the uncertainty faced by the public under alternative monetary arrangements when monetary control is imperfect. The variance of the monetary control error is the same in each regime. We compare three alternative monetary arrangements: policy is discretionary and there are no announcements of targets; policy is discretionary but the policymaker is required to announce a target, as in the United States since 1975; there is a rule of the type proposed by Friedman (1960) that requires constant monetary growth. Uncertainty is measured by the variance of the one-period-ahead error in predicting monetary growth.

We begin by considering the effect of announcements. The variances of the one-period-ahead errors in the presence and in the absence of announced targets are respectively

$$V^*(e) \equiv E[m_i - E[m_i|I_i^*]]^2 \qquad (21a)$$

$$V(e) = E[m_i - E[m_i|I_i]]^2 \qquad (21b)$$

where I_i^* is the public's information set in the presence of announcements and I_i is its information set in the absence of announcements. In this section we will identify parameters and variables in the economy with announcements by a starred superscript. Thus B^* now denotes the coefficient of p_i in equation (7) while the symbol B (previously used to denote this coefficient) is reserved for the same coefficient in the absence of announcements.

The explicit form of $B^*(\cdot)$ is (from equation (16))

$$B^*(\cdot) \equiv \frac{(1 - \beta\rho^2)\delta(1 + \sigma_\psi^2/\sigma_\gamma^2)}{(1 - \rho\beta\delta)(\delta + \rho\sigma_\psi^2/\sigma_\gamma^2)} \qquad (22)$$

The explicit form of $B(\cdot)$ is (from equation (16) of CM)

$$B = \frac{1 - \beta\rho^2}{1 - \beta\rho\lambda} \qquad (23)$$

where λ is the value to which δ tends when σ_γ^2 tends to infinity. Using (8a) and (9) in (21a), using equations (7) and (9) from CM in (21b), and rearranging we obtain expressions for the variance of unanticipated monetary growth with and without announcements.[18]

$$V^*(e) = E[\psi_i + B^*(p_i - E[p_i|z_i, z_{i-1},\ldots, y^*_{i-1}, y^*_{i-2},\ldots])]^2 \quad (24a)$$

$$V(e) = E[\psi_i + B(p_i - E[p_i|y_{i-1}, y_{i-2},\ldots])]^2 \quad (24b)$$

where

$$y^*_i = B^* p_i + \psi_i \quad (25)$$

Noting that ψ_i is distributed independently of its lagged values and of γ and using the fact that the conditional expected values of p_i are minimum mean square error estimators of p_i, equations (24a) and (24b) can be rewritten

$$V^*(e) = \sigma^2_\psi + [B^*(\sigma^2_\psi, \sigma^2_\gamma)]^2 \min_{\{a^*_j, c_j\}} E\left[p_i - \sum_{j=1}^\infty a^*_j y^*_{i-j} - \sum_{j=0}^\infty c_j z_{i-j}\right]^2 \quad (26a)$$

$$V(e) = \sigma^2_\psi + [B(\sigma^2_\psi)]^2 \min_{\{a_j\}} E\left[p_i - \sum_{j=1}^\infty a_j y_{i-j}\right]^2 \quad (26b)$$

Here a^*_j and c_j are, respectively, the weights of y^*_{i-j} and of z_{i-j} in the expression for the optimal predictor of p_i, and a_j is the weight of y_{i-j} when this predictor is based solely on observations of y. The explicit form of those weights is given, respectively, by equation (10a) of this article and by equation (10a) of CM.

Passing $B^*(\cdot)$ and $B(\cdot)$ in equations (26a) and (26b) to the right of the min operators, redefining variables, and rearranging, we obtain

$$V^*(e) = \sigma^2_\psi + \min_{\{a^*_j, c_j\}} E\left[q^*_i - \sum_{j=1}^\infty a^*_j(q^*_{i-j} + \psi_{i-j}) - \sum_{j=0}^\infty c_j(q^*_{i-j} + \gamma_{i-j})\right]^2 \quad (27a)$$

$$V(e) = \sigma^2_\psi + \min_{\{a_j\}} E\left[q_i - \sum_{j=1}^\infty a_j(q_{i-j} + \psi_{i-j})\right]^2 \quad (27b)$$

where

$$q^*_j \equiv B^* p_j, \quad q_j \equiv B p_j, \quad \text{for all } j \quad (28)$$

The definitions in (28) in conjunction with (5) imply

$$Eq = Eq^* = 0, \quad \sigma^2_q = B^2 \sigma^2_p, \quad \sigma^2_{q^*} = (B^*)^2 \sigma^2_p,$$

$$Eq_i q_{i-j} = \rho^j \sigma^2_q, \quad Eq^*_i q^*_{i-j} = \rho^j \sigma^2_{q^*} \quad \text{for all } i \text{ and } j \quad (29)$$

---

18. Instead of reestablishing all of the results used in the case of no announcements, we refer the reader to the relevant sections of our earlier paper, Cukierman and Meltzer (1986), reprinted here as chapter 8.

where σ_q^2 and $\sigma_{q^*}^2$ are the variances of q and q^*, respectively.

Expanding the square terms on the right-hand sides of equations (27a) and (27b), using (29) and the fact that ψ_j^*, ψ_j, and γ_j are all mutually and serially uncorrelated, equations (27) can be rewritten

$$V^*(e) = \sigma_\psi^2 + \min_{\{a_j^*, c_j\}} \left\{ \sum_{j=1}^\infty (a_j^*)^2 \sigma_\psi^2 + \sum_{j=0}^\infty c_j^2 \sigma_\gamma^2 \right.$$

$$+ \left[(1-c_0)^2 - 2(1-c_0) \sum_{j=1}^\infty (a_j^* + c_j)\rho^j \right.$$

$$\left. \left. + \sum_{j=1}^\infty \sum_{t=1}^\infty (a_j^* + c_t)(a_t^* + c_t)\rho^{|j-t|} \right] \sigma_{q^*}^2 \right\} \tag{30a}$$

$$V(e) = \sigma_\psi^2 + \min_{\{a_j\}} \left[\sum_{j=1}^\infty a_j^2 \sigma_\psi^2 + \left(1 - 2\sum_{j=1}^\infty a_j \rho^j \right. \right.$$

$$\left. \left. + \sum_{j=1}^\infty \sum_{t=1}^\infty a_j a_t \rho^{|j-t|} \right) \sigma_q^2 \right] \tag{30b}$$

If $\sigma_{q^*}^2$ and σ_q^2 are equal, the minimum on the right-hand side of (30a) is no larger than the minimum on the right-hand side of (30b). That is,

$$V^*(e, \sigma_{q^*}^2) \leqslant V(e, \sigma_q^2) \equiv V(e). \tag{31}$$

The reason is that the value of the minimum in (30b) can always be attained in (30a) by setting the minimizers in (30a) at the values $a_j^* = a_j$, $j = 1, 2, \ldots$ and $c_j = 0$, $j \geqslant 0$. But in fact

$$\sigma_{q^*}^2 \equiv [B^*(\sigma_\psi^2, \sigma_\gamma^2)]^2 \sigma_p^2 < [B(\sigma_\psi^2)]^2 \sigma_p^2 \equiv \sigma_q^2 \tag{32}$$

because $B^*(\cdot) < B(\cdot)$.[19] By the envelope theorem, the total effect of a change in $\sigma_{q^*}^2$ on $V^*(e)$ is equal to the direct effect of $\sigma_{q^*}^2$ on $V^*(e)$ evaluated at the minimizing values of $\{a_j^*, c_j\}$. This partial derivative is given in turn by the coefficient of $\sigma_{q^*}^2$ in (30a) evaluated at the minimum. For any $\sigma_{q^*}^2$, and for $\sigma_{q^*}^2 = \sigma_q^2$ in particular, this coefficient is positive provided $\rho < 1$.[20] It follows from this observation and (32) that

$$V^*(e) \equiv V^*(e, \sigma_{q^*}^2) < V^*(e, \sigma_q^2) \tag{33}$$

This, together with (31), implies that preannouncement lowers the variance of the forecast errors, that is

$$V^*(e) < V(e), \tag{34}$$

---

19. See part 3 of the appendix for a proof.
20. The proof appears in part 4 of the appendix.

Friedman (1960) proposes a rule requiring a constant growth of money. The policymaker in our model cannot control money growth without error. A rule for constant, preannounced money growth sets a constant value of $m_i^p = m_i^a$ but does not make m_i constant because of the control error. The variance of the one-period-ahead forecast error is, in this case, due entirely to imperfect control of the money supply and is given by σ_ψ^2. It follows from this observation in conjunction with (27) and (34) that

$$\sigma_\psi^2 < V^*(e) < V(e) \tag{35}$$

Equation (35) is the main result of this section. It states that, for an exogenously given variance of the monetary control error, announcements reduce uncertainty and that uncertainty can be further reduced if the policymaker follows a Friedman-type rule. This result suggests that as long as σ_γ^2 is not infinite the requirement to preannounce targets is not meaningless. When the policymaker behaves in a discretionary manner, mandatory announcements of targets reduce uncertainty for the public. Abandoning discretion by adopting Friedman's rule further reduces uncertainty.

In the absence of announcements, the level of monetary uncertainty imposed on the public rises with the policymaker's time preference[21] (proposition 3 of CM). Equations (30a) and (30b) also imply that the higher the variance of the monetary control error, σ_ψ^2, the higher the level of monetary uncertainty imposed on the public both in the presence and in the absence of preannouncements. The level of uncertainty in the presence of announcements is also larger the larger the noisiness of those announcements as measured by σ_γ^2. Both results are direct consequences of the envelope theorem and of the fact that the coefficients of σ_ψ^2 and of σ_γ^2 in equations (30a) and (30b) are positive.

We close the discussion of uncertainty by relating the reduction of uncertainty achieved by Friedman's rule to the credibility problem. The reduced level of uncertainty achieved by a rule requiring a constant planned rate of money growth can be achieved under discretion and announcements if the noise component of those announcements has zero variance, that is, if $\sigma_\gamma^2 = 0$. We showed above that in this case (11) implies that $E[m_i|I_i] = m_i^a$. From (6), $m_i^a = m_i^p$. The forecast error is $m_i - E[m_i|I_i^*] = \psi_i$ with variance σ_ψ^2. Hence the level of uncertainty can be reduced to that of Friedman's rule even if planned money growth is not required to be constant, provided announcements have perfect credibility. But when $\sigma_\gamma^2 = 0$, the public gets advance perfect information about any changes in the plans of the policymaker, and the policymaker's information advantage disappears. There is no benefit to the policymaker from planning any money growth rate other than zero. Perfectly credible announcements induce the policymaker to adhere to a Friedman-type rule even when he has the discretion not to do so.

---

21. It is likely that the lower the value of β for given values of the noise variances σ_ψ^2 and σ_γ^2 the higher the variance of the mean forecast error in the presence of preannouncements. We have not proved this, however.

The variability of monetary growth under the three monetary arrangements has the same ordering as uncertainty. Friedman's monetary rule generates the lowest variability; discretionary policy without announcements generates the most. By comparing (18) in CM with (17),

$$V(m_i) = \sigma_\psi^2 + [B(\sigma_\psi^2)]^2 \sigma_p^2 \tag{36a}$$

$$V^*(m_i) = \sigma_\psi^2 + [B^*(\sigma_\psi^2, \sigma_\gamma^2)]^2 \sigma_p^2 \tag{36b}$$

Under Friedman's rule, m_i^p is constant, so the variance of monetary growth with such a rule is σ_ψ^2. σ_ψ^2 is lower than the variances in (36). Further, for a given variance of the control error, announcements reduce the variance of monetary growth. This is a direct consequence of (36) in conjunction with the fact[22] that $B^*(\sigma_\psi^2, \sigma_\gamma^2) < B(\sigma_\psi^2)$. The positive implication of this discussion is that, other things the same, preannouncements reduce the variability of monetary growth.

VI. Concluding Comments

This chapter has developed a positive theory of credibility for a monetary system in which control is imperfect and announced targets are inaccurate. The level of credibility varies inversely with the difference between the announcement and the public's expectations of money growth. Credibility is perfect when the announced and the expected rates of monetary growth are identical.

This concept of credibility seems appropriate for countries like the United States in which policymakers operate in a discretionary manner but are required to make announcements about the growth of monetary aggregates. In contrast, Taylor (1982) and Barro and Gordon (1983a) focused on a somewhat different concept of credibility. They studied the conditions that lead people to believe that the government has abandoned discretion in favor of a rule mandating a zero rate of money growth.

Our analysis suggests that the legal requirement to announce targets does not generate immediate credibility if announcements are noisy. In such cases credibility of new policies is established slowly. Given the precision of monetary control, however, preannouncements almost always reduce the level of monetary uncertainty faced by the public.

The credibility of newly instituted disinflationary policies also depends on the quality of monetary control. With tight control, a few periods of determined slowdown in the rate of monetary expansion, accompanied by accurate announcements of the new policy, suffice to convince the public that money growth is permanently lower. As a consequence, expectations of inflation fall quickly. Unexpected rates of monetary growth remain negative for a relatively brief period, and the accompanying unemployment is relatively small. In this case a "cold turkey" disinflationary policy is preferable to "gradualism," since a

---

22. See part 3 of the appendix.

large decrease in monetary expansion, preannounced to the public, generates credibility relatively quickly. If the policymaker has poor control of money growth and is inaccurate in his announcements, disinflationary policy takes substantially longer to become credible, however. The interim period of unemployment is longer and unemployment is larger. The costs of disinflation are higher. A gradual approach, which permits the public to adjust anticipations and credibility, seems preferable in these circumstances.

Fischer (1986) provided estimates of the costs of disinflation. His analysis suggests that those costs are quite sensitive to the way expectations adjust. Our chapter relates this critical speed of adjustment to some underlying factors like the quality of monetary control, the quality of monetary announcements, and the discount factor of the policymaker. Fischer's analysis in conjunction with ours creates a link between the costs of disinflation on the one hand and the quality of monetary control and the degree of time preference of the policymaker on the other.

In Cukierman and Meltzer (1986) we analyzed the circumstances under which the policymaker will choose imprecise control procedures even if perfect control is technically feasible. Given such a choice, any additional information about the current objectives of the policymaker is potentially valuable to the public. Preannouncement of monetary targets is one type of additional, imprecise but not meaningless, information. Statements made by Fed officials about their view of the current state of the economy and related issues are another. The analysis here may be viewed more generally as applying to any kind of signal issued by the central bank about its future monetary policy besides past money growth. With this wider interpretation in mind, the chapter implies that the lower the precision of monetary control, the more attention will be paid to other signals of planned monetary growth like public statements, rumors, and personalities. In such circumstances any public appearance by a high-ranking Fed official receives wide press coverage. With very precise monetary control, on the other hand, past money growth is a sufficient statistic for future plans, and the pronouncements of central bank officials are given little attention.

Appendix

1. *Derivation of the optimal predictors in equation* (10) *of the text.*

Define the stochastic variables $\varepsilon_t \equiv \psi_t/B$ and $\eta_t \equiv \gamma_t/B$. Substituting those relations into equations (8a) and (8b)

$$y_t = B(p_t + \varepsilon_t) \equiv By_t^1; \qquad z_t = B(p_t + \eta_t) \equiv Bz_t^1 \tag{A1}$$

Since the public knows the parameter B, an observation on the pair (y_t, z_t) is equivalent to an observation on the pair (y_t^1, z_t^1). Hence the expected value of p_t conditioned on past values of y and z is equal to this expected value conditioned on past values of y^1 and z^1. We turn next to the calculation of this expected value.

Since p_t, y_t^1, and z_t^1 are all normally distributed, this expected value is a linear function with fixed coefficients of the observations on past values of y^1 and z^1. That is

$$E[p_t | I_t^1] = \sum_{i=1}^{\infty} a_i y_{t-i}^1 + \sum_{i=0}^{\infty} c_i z_{t-i}^1 \tag{A2}$$

where

$$I_t^1 \equiv \{y_{t-1}^1, y_{t-2}^1, \ldots, z_t^1, z_{t-1}^1, z_{t-2}^1, \ldots\}$$

Since the conditional expected value in (A2) is also the point estimate of p_t, which minimizes the mean square error around this estimate, it follows that $\{a_i\}_{i=1}^{\infty}$, $\{c_i\}_{i=0}^{\infty}$ are to be chosen so as to minimize

$$Q \equiv E\left[p_t - \sum_{i=1}^{\infty} a_i y_{t-i}^1 - \sum_{i=0}^{\infty} c_i z_{t-i}^1\right]^2$$

Using (A1) in this expression, using the fact that ε, η, and v are mutually independent, and passing the expectation operator through, Q can be rewritten

$$Q = \{(1 - c_0)^2 + [(1 - c_0)\rho - (a_1 + c_1)]^2$$
$$+ [(1 - c_0)\rho^2 - \rho(a_1 + c_1) - (a_2 + c_2)]^2$$
$$+ \cdots + [(1 - c_0)\rho^i - \rho^{i-1}(a_1 + c_1) - \cdots - (a_i + c_i)]^2 + \cdots\}\sigma_v^2$$
$$+ \sum_{i=1}^{\infty} a_i^2 \sigma_\varepsilon^2 + \sum_{i=0}^{\infty} c_i^2 \sigma_\eta^2 \tag{A3}$$

where $\sigma_\varepsilon^2 = \sigma_\psi^2/B^2$ and $\sigma_\eta^2 = \sigma_\gamma^2/B^2$.

The necessary first-order conditions for an extremum of Q are[23]

$$\frac{\partial Q}{\partial c_0} = -2\{1 - c_0 + \rho[(1 - c_0)\rho - (a_1 + c_1)] + \cdots$$
$$+ \rho^i[(1 - c_0)\rho^i - \rho^{i-1}(a_1 + c_1) - \cdots - (a_i + c_i)] + \cdots\}\sigma_v^2$$
$$+ 2c_0 \sigma_\eta^2 = 0 \tag{A4a}$$

$$\frac{\partial Q}{\partial c_i} = -2\{[(1 - c_0)\rho^i - \rho^{i-1}(a_1 + c_1) - \cdots - (a_i + c_i)]$$
$$+ \rho[(1 - c_0)\rho^{i+1} - \rho^i(a_1 + c_1) - \cdots - (a_{i+1} + c_{i+1})] + \cdots\}\sigma_v^2$$
$$+ 2c_i \sigma_\eta^2 = 0 \quad i \geq 1 \tag{A4b}$$

23. Since Q is a quadratic, this extremum is a minimum.

$$\frac{\partial Q}{\partial a_i} = -2\{[(1-c_0)\rho^i - \rho^{i-1}(a_1+c_1) - \cdots - (a_i+c_i)]$$
$$+ \rho[(1-c_0) - \rho^{i+1} - \rho^i(a_1+c_1) - \cdots - (a_{i+1}+c_{i+1})] + \cdots\}\sigma_v^2$$
$$+ 2a_i\sigma_\varepsilon^2 = 0 \quad i \geq 1. \tag{A4c}$$

Leading (A4b) by one period, multiplying by ρ, and subtracting (A4b) from the resulting expression

$$[(1-c_0)\rho^i - \rho^{i-1}(a_1+c_1) - \cdots - (a_i+c_i)]\sigma_v^2 + (\rho c_{i+1} - c_i)\sigma_\eta^2 = 0 \quad i \geq 1. \tag{A5}$$

Multiplying (A5) by ρ, subtracting the resulting expression from (A5) led by one period, and rearranging

$$\rho\sigma_\eta^2 c_{i+2} - (\sigma_v^2 + (1+\rho^2)\sigma_\eta^2)c_{i+1} - \sigma_v^2 a_{i+1} + \rho\sigma_\eta^2 c_i = 0 \quad i \geq 1. \tag{A6}$$

Rearranging (A4b) and (A4c), equating their common terms, and rearranging again

$$a_i = \frac{\sigma_\eta^2}{\sigma_\varepsilon^2} c_i \quad i \geq 1. \tag{A7}$$

Substituting (A7) into (A6) and dividing the resulting expression by $\rho\sigma_\eta^2$

$$c_{i+2} - \left\{\frac{1+r}{\rho} + \rho\right\}c_{i+1} + c_i = 0 \quad i \geq 1 \tag{A8}$$

where

$$r \equiv \frac{\sigma_v^2}{\sigma_\varepsilon^2}\left(1 + \frac{\sigma_\varepsilon^2}{\sigma_\eta^2}\right) = B^2 \frac{\sigma_v^2}{\sigma_\psi^2}\left(1 + \frac{\sigma_\psi^2}{\sigma_\gamma^2}\right) \tag{A9}$$

(A8) is a second-order homogeneous difference equation whose general solution is

$$c_i = K\delta^i \quad i \geq 1 \tag{A10}$$

where K is a constant to be determined by initial conditions and δ is the root of the quadratic

$$u^2 - \left(\frac{1+r}{\rho} + \rho\right)u + 1 = 0 \tag{A11}$$

The roots of this equation are given by

$$u_{1,2} = \frac{1}{2}\left(\frac{1+r}{\rho} + \rho\right) \pm \sqrt{\left\{\frac{1}{2}\left(\frac{1+r}{\rho} + \rho\right)\right\}^2 - 1}. \quad (A12)$$

The positive root in (A12) is larger than one and the negative root is bounded between zero and one. Thus c_i does not diverge only if the smaller root is substituted for δ in (A10). Since c_i has to yield a minimum for Q, it cannot diverge. Hence

$$\delta = \frac{1}{2}\left(\frac{1+r}{\rho} + \rho\right) - \sqrt{\left\{\frac{1}{2}\left(\frac{1+r}{\rho} + \rho\right)\right\}^2 - 1}. \quad (A13)$$

Substituting (A10) into (A7)

$$a_i = K\frac{\sigma_\eta^2}{\sigma_\varepsilon^2}\delta^i \qquad i \geq 1 \quad (A14)$$

K is determined from the initial conditions. Substitute (A10) and (A14) into (A5) for the case $i = 1$ and rearrange to get

$$\rho\sigma_v^2(1 - c_0) - \sigma_\eta^2(1 + r)K\delta + \rho\sigma_\eta^2 K\delta^2 = 0. \quad (A15)$$

Specializing (A4b) to the case $i = 1$, multiplying by ρ, subtracting (A4a) from the resulting expression, and rearranging

$$c_0 = \frac{\sigma_v^2}{\sigma_v^2 + \sigma_\eta^2} + \frac{\sigma_\eta^2}{\sigma_v^2 + \sigma_\eta^2}\rho c_1. \quad (A16)$$

Substituting (A16) into (A15), using (A10), and rearranging

$$K = \frac{\sigma_v^2}{\sigma_v^2 + \sigma_\eta^2(1 - \rho\delta)} \quad (A17)$$

Multiplying numerator and denominator of (A17) by $(\sigma_\varepsilon^2 + \sigma_\eta^2)/\sigma_\eta^2\sigma_\varepsilon^2$ and noting the definition of r in (A9)

$$K = \frac{r}{r + 1 - \rho\delta + (\sigma_\eta^2/\sigma_\varepsilon^2)(1 - \rho\delta)}. \quad (A17a)$$

Since δ is a root of (A11) it satisfies this equation, that is,

$$\delta^2 - \left(\frac{1+r}{\rho} + \rho\right)\delta + 1 = 0. \quad (A18)$$

Solving for r from (A18), substituting into (A17a), and rearranging it follows that

$$K = \frac{\sigma_\varepsilon^2(\rho - \delta)}{\rho\sigma_\varepsilon^2 + \delta\sigma_\eta^2} \tag{A17b}$$

Substituting K from (A17b) into (A10) and (A14)

$$c_i = \frac{(\rho - \delta)(\sigma_\varepsilon^2 + \sigma_\eta^2)}{\rho\sigma_\varepsilon^2 + \delta\sigma_\eta^2}(1 - \theta)\delta^i \qquad i \geq 1 \tag{A19a}$$

$$a_i = \frac{(\rho - \delta)(\sigma_\varepsilon^2 + \sigma_\eta^2)}{\rho\sigma_\varepsilon^2 + \delta\sigma_\eta^2}\theta\delta^i \qquad i \geq 1 \tag{A19b}$$

where

$$\theta \equiv \frac{\sigma_\eta^2}{\sigma_\varepsilon^2 + \sigma_\eta^2} = \frac{\sigma_\gamma^2}{\sigma_\psi^2 + \sigma_\gamma^2}.$$

Substituting (A17) into (A10) specializing to the case $i = 1$, substituting the resulting expression into (A16), and rearranging

$$c_0 = \frac{\sigma_v^2}{\sigma_v^2 + (1 - \rho\delta)\sigma_\eta^2} = \frac{(\rho - \delta)(\sigma_\varepsilon^2 + \sigma_\eta^2)}{\rho\sigma_\varepsilon^2 + \delta\sigma_\eta^2}(1 - \theta) \tag{A20}$$

The second equality in (A20) follows by noting that K is given by both (A17) and (A17b).

The optimal predictor is obtained by inserting equations (A19) and (A20) into (A1) and is given by

$$E[p_t|I_t] \equiv \frac{(\rho - \delta)(\sigma_\varepsilon^2 + \sigma_\eta^2)}{\rho\sigma_\varepsilon^2 + \delta\sigma_\eta^2} \left\{ \sum_{j=1}^{\infty} \delta^j(\theta y_{t-j}^1 + (1 - \theta)z_{t-j}^1) + (1 - \theta)z_t^1 \right\}. \tag{A21}$$

Equation (10) in the text follows by using (A1) and the fact that $\sigma_\varepsilon^2 = \sigma_\psi^2/B^2$, $\sigma_\eta^2 = \sigma_\gamma^2/B^2$ in (A21).

2. Derivation of Equation (16)

Substituting (4) and (15) into (14) and rearranging the resulting expression we obtain

$$m_i^p = \frac{\delta}{(\rho - \delta)(1 - \theta) + \delta}[A + p_i - (\rho - \delta)(\rho\beta + (\rho\beta)^2\delta + (\rho\beta)^3\delta^2 + \cdots)(A + p_i)$$
$$- (\rho - \delta)\{(\beta + \beta^2\delta + \beta^3\delta^2 + \cdots) - [\rho\beta + (\rho\beta)^2\delta + (\rho\beta)^3\delta^2 + \cdots]\}A]. \tag{A22}$$

Noting that

$$(\rho - \delta)(1 - \theta) + \delta = \frac{\delta\sigma_\eta^2 + \rho\sigma_\varepsilon^2}{\sigma_\eta^2 + \sigma_\varepsilon^2}$$

summing the geometric progressions in (A22) and collecting terms

$$m_i^p = \frac{\delta(\sigma_\eta^2 + \sigma_\varepsilon^2)}{\sigma_\eta^2 \delta + \sigma_\varepsilon^2 \rho} \left\{ \left[1 - \frac{(\rho - \delta)\beta}{1 - \beta\delta} \right] A + \left[1 - \frac{(\rho - \delta)\rho\beta}{1 - \rho\beta\delta} \right] p_i \right\}. \quad (A23)$$

(A23) can be rewritten

$$m_i^p = \frac{\delta(1 - \rho\beta)(\sigma_\eta^2 + \sigma_\varepsilon^2)}{(1 - \delta\beta)(\delta\sigma_\eta^2 + \rho\sigma_\varepsilon^2)} A + \frac{\delta(1 - \beta\rho^2)(\sigma_\eta^2 + \sigma_\varepsilon^2)}{(1 - \rho\beta\delta)(\delta\sigma_\eta^2 + \rho\sigma_\varepsilon^2)} p_i. \quad (A24)$$

Equation (16) in the text follows by dividing both numerators and denominators in (A24) by σ_η^2 and using the identities in (A1).

3. *Proof that* $B^*(\sigma_\psi^2, \sigma_\gamma^2) < B(\sigma_\psi^2)$

From equations (23) and (16)

$$B(\cdot) = \frac{1 - \beta\rho^2}{1 - \beta\rho\lambda}; \quad B^*(\cdot) = \frac{1 - \beta\rho^2}{1 - \beta\rho\delta} \cdot \frac{\delta(1 + d)}{\delta + \rho d} \quad (A25)$$

where $d \equiv \sigma_\psi^2/\sigma_\gamma^2$.

Since $\rho \geq \delta$, $\delta(1 + d)/(\delta + \rho d) \leq 1$.

Note (using (10c) in CM and (10c) of this chapter) that for any finite σ_γ^2, r is larger in the presence of announcements. Since δ is a decreasing function of r, it follows that for given σ_ψ^2 and σ_γ^2

$$\delta < \lambda$$

Then, together with (A26) and the fact that $\delta(1 + d)/(\delta + \rho d) \leq 1$, implies that

$$B^*(\sigma_\psi^2, \sigma_\gamma^2) < B(\sigma_\psi^2)$$

4. *Proof that the coefficients of $\sigma_{q^*}^2$ and σ_q^2 in equations (30a) and (30b) are positive at the minimum for any $\rho < 1$.*

The difference between the minimization problem in (26a) and in (27a) is only that in the latter equation the constant B^* has been absorbed into the minimization problem. Hence the minimizing values for the problem in (27a) are the same as those for the problem in (26a) for which the minimizing values of $\{a_j^*\}$ and $\{c_j\}$ have been derived in part 1 of the appendix and are given by

$$a_j^* = \frac{(\rho - \delta)\theta}{\rho(1 - \theta) + \delta\theta} \delta^j, j = 1, 2, \ldots;$$

$$c_j = \frac{(\rho - \delta)(1 - \theta)}{\rho(1 - \theta) + \delta\theta} \delta^j, j = 0, 1, \ldots \quad \text{(A26)}$$

(Those values of a^* are called a in part 1 of the appendix. The use of the asterisk is consistent with the notation in Section V of the text).

Substituting (A26) into the coefficient of $\sigma_{q^*}^2$ in (30a) and using repeatedly the formula for the summation of infinite geometric progressions, we obtain after a considerable amount of algebra the following expression for the effect of $\sigma_{q^*}^2$ on $V^*(e)$

$$\frac{\partial V^*(e)}{\partial \sigma_{q^*}^2} = \frac{\delta^2 (1 - \rho^2)}{[\rho(1 - \theta) + \delta\theta]^2 (1 - \delta^2)}$$

This expression is unambiguously positive for any $\rho < 1$. Since the proof holds for any positive value of $\sigma_{q^*}^2$, including in particular all the range between $\sigma_{q^*}^2$ and σ_q^2, it follows that

$$\frac{\partial V^*(e)}{\partial \sigma_{q^*}^2} > 0$$

in all the range $[\sigma_{q^*}^2, \sigma_q^2]$.

The proof that $\partial V(e)/\partial \sigma_q^2 > 0$ for all $\rho < 1$ proceeds along analogous lines.

References

Barro, R. J. and Gordon, D. B. (1983a). "Rules, Discretion and Reputation in a Model of Monetary Policy." *Journal of Monetary Economics, 12* (July), pp. 101–22.

Barro, R. J and Gordon, D. B. (1983b). "A Positive Theory of Monetary Policy in a Natural Rate Model." *Journal of Political Economy, 91* (Aug.), pp. 589–610.

Beck, N. (1982). "Presidential Influence on the Federal Reserve in the 1970s." *American Journal of Political Science, 26* (Aug.), pp. 415–45.

Brunner, K. and Meltzer, A. H. (1983). "Strategies and Tactics for Monetary Control." *Carnegie-Rochester Conference on Public Policy, 18* (Spring), pp. 59–104.

Cukierman, A. (1984). *Inflation, Stagflation, Relative Prices and Imperfect Information.* Cambridge: Cambridge University Press.

Cukierman, A. (1986). "Central Bank Behavior and Credibility—Some Recent Theoretical Developments." *Federal Reserve Bank of St. Louis Review, 68* (May), pp. 5–17.

Cukierman, A. and Meltzer, A. H. (1986). "A Theory of Ambiguity, Credibility and Inflation under Discretion and Asymmetric Information." *Econometrica*, 54 (Sep.), pp. 1099–1128.

Fischer, S. (1986). "Contracts, Credibility and Disinflation." In *Indexing, Inflation and Economic Policy*, Cambridge, Mass.: MIT Press. pp. 221–46.

Friedman, M. (1960). *A Program for Monetary Stability*. New York: Fordham University Press.

Hetzel, R. L. (1984). "The formulation of Monetary Policy in a Democracy." Unpublished Manuscript, Federal Reserve Bank of Richmond, January.

Kane, E. J. (1980). "Politics and Fed Policymaking: The More Things Change, the More They Remain the Same." *Journal of Monetary Economics*, 6 (Apr.), pp. 199–212.

Kane, E. J. (1980), "External Pressure and the Operation of the Fed." In *Political Economy of International and Domestic Monetary Relations*, R. E. Lombra and W. E. Witte, eds. Ames: Iowa University Press, pp. 211–32.

Kydland, F. E. and Prescott, E. C. (1977). "Rules Rather than Discretion: The Inconsistency of Optimal Plans." *Journal of Political Economy*, 85 (June), pp. 473–92.

Taylor, J. B. (1982). "Establishing Credibility: A Rational Expectations Viewpoint." *American Economic Review*, 72 (May), pp. 81–85.

Weintraub, R. E. (1978). "Congressional Supervision of Monetary Policy." *Journal of Monetary Economics*, 4 (Apr.), pp. 341–62.

Wooley, J. T. (1984). *The Federal Reserve and the Politics of Monetary Policy*. Cambridge, Cambridge University Press.

Index

Ability
  income and, 82
  progressive tax structure and distribution of, 96–97
Abortion, 12
Activist policy, differences in preference for, 151–54, 155
Aiyagari, R. S., 132n.28
Alchian, A. A., 8
Alt, J., 31, 44
Altig, D., 118n.9
Ambiguity
  costs arising from, 16
  politically optimal level of, 158–59, 174–80, 181–82, 187–88
Anderson, M., 54, 55n.6
Aranson, P. H., 24, 37n.1, 42n.8
Arrow, K. J., 79, 85
Asymmetric information, 142–45, 148, 155, 170, 171, 175
Atkinson, A. B., 55, 77
Authoritarian government, 14–15
Average credibility, 191–92, 199–200
Average wage rate, 112

Backus, D., 172, 173, 189
Barro, R. J., 19, 109, 110, 111, 119, 121, 132, 159, 160, 170–72, 192n.3, 205
Beck, N., 193
Becker, G., 7, 11n.6
Bedand, R., 53n.1
Beliefs of public, expected rate of monetary growth and, 164–65
Bequest-contained individuals, 19, 110–11, 128n.23
  attributes of, 114–18

  benefits from intergenerational reallocation of taxes through debt, 119–20
  debt determination in absence of general equilibrium effects, 122–23
  effects of debt on welfare of, 126–27
  likelihood of voting for higher debt, 129
Bequest motive, operative, 109–10, 119
Bequests
  change in welfare induced by realignment in, 124, 125
  inter vivos, 118n.9
  the poor and, 19
Bernanke, B. S., 136n.2, *156*
Bhansali, R. J., 161, *189*
Bias
  toward economic stimulation, 169
  inflationary, 21, 169, 170
  simultaneous-equations, 43
Blum, W., 18, 76
Bomhoff, E. J., 163n.7
Borcherding, T., 56
Bös, D., 18n.13
Brennan, G., 6
Browning, E. K., 16n.11, 26n.5
Brunner, K., 8, 24, 159n.1, 163n.7, 194n.7
Buchanan, J. M., 5, 6, 7, 10n.5, 24n.3, 33
Budget, balanced, 24, 40
Budget deficits. *See* Deficits
Buiter, W. H., 113n.4
Bull, C., 160, 172
Burbidge, J. B., 113n.4
Burns, A. F., 158, 162
Bush, W. C., 6

Cameron, D., 24, 32, 36, 44
Canzoneri, M. B., 139, 173n.21

Index

Capital
  crowding out of, 110–11, 120–28
  as substitute for debt in production vs. portfolios, 110
Carmichael, J., 113n.4
Cash-equivalent mix, 56, 59, 61–62, 73–74
Cash-equivalent transfer, 54, 56
Cash payments, 56
Central banks, 20, 158, 159. *See also* Federal Reserve; Monetary announcements; Monetary growth
  model of behavior of, 193–97
  monitoring of, 172
  statutory independence of, 193
Change(s)
  coercion and, 9–11
  democratic vs. nondemocratic, 4–5
  in memory, 20–21
Children, investment in human capital of, 133
Coercion, change and, 9–11
Congress, 54, 158
Constitutional rules, 6–7, 20, 137, 146–48, 154, 173
Constitutions, 4–9
Consumption, 27
  by the bequest-constrained vs. non-bequest-constrained, 121
  intergenerational reallocation of, 123–24, 125
  marginal utility of, 81
  as normal goods, 39
Consumption-leisure relationship, utility function for, 26, 38–39
Consumption tax, linear, 54–55
Contractarian position, 4–9, 11, 12–13
Cooter, R., 32n.11, 42n.8
Corlett, W. J., 55
Cost(s)
  arising from ambiguity and credibility of policy, 16
  of democracy, 20, 136, 137, 145, 146, 148, 154
  of disinflation, 181, 200, 206
Credibility
  concepts of, 173
  cost of, 16
  defined, 20, 168
  determinants of, 168
  distribution of monetary growth and, 159, 168–70
  imperfect, 173–74
  level of noise in money control and, 179, 182
  of monetary announcements, 20, 199–201
    determinants of, 199–201; measures of, 191–92, 199–200; model for analyzing, 193–97; uncertainty under alternative monetary arrangements and, 201–5
  of policymakers, 20–21
  speed of public learning and, 182
Credibility hypothesis, 182

Crowding out of capital, 110–11, 120–28
Cukierman, A., 103, 116, 120n.12, 127n.22, 130, 132, 136n.2, 163n.7, 169, 181n.31, 192n.3, 4, 193, 202n.18, 206

Davis, S. J., 118n.9
Debt
  coalitions favoring and opposing larger, 129–30
  decisive voter and, 33
  determination of, by majority rule, 120–28
  effects on welfare of bequest- vs. non-bequest-constrained individuals, 125–27
  financing of, general equilibrium effects of, 122
  individual benefits from intergenerational reallocation of taxes through, 118, 119–20
  political theory of, 109–34
  as substitute for capital in production vs. portfolios, 110
Debt neutrality theorem, 110
Decision making, individual-collective relationship in, 16, 33
Decisive voter, 29
  government debt and, 33
  government size and, 25, 31, 32–33, 37, 40–42, 46, 70
  income of
    earned, 41; mean income relative to, 25, 31, 37
  income tax choices, 60–64, 67–69, 71–72
    by nonworking voter, 99–100; optimal, 60–64, 70
    by working voter, 95–99
  median income voter as, 46, 60, 79
  productivity of, 40–41, 72
  reasoning about redistribution, 70–71
  tax scheduling choices of, 83–95
    globally winning schedule, 91–95; majority-winning schedule, 85–91
  voting rule and, 70, 71
Defense, military draft for, 148–50
Deficits, 19. *See also* Debt
  coalitions favoring and opposing, 129–30
  economic conditions conducive to, 131–32
  during wars, 132
Democracy
  costs of, 20, 136, 137, 145, 146, 148, 154
  liberal society vs., 9–10
Democratic vs. nondemocratic change, 4–5
Developing countries, income distribution in, 14–15
Diamond, P. A., 55, 111
Dictatorship, 29
Discount factor, optimal level of ambiguity and, 179–80
Discretion, rules vs., 19–20, 135
Discretionary policy, 135–57, 172–73
  constitutional rules, implementation of, 137, 146–48, 154
  example of, 148–51

Index

politically motivated government and
   facing voters with diverse objectives, 137, 151–54, 155; with fully informed voters, 146; with partially informed voters, 136, 142–45
  social planner's problem in, 138–41
  uncertainty and variability of monetary growth under, 201–3, 205
Disinflationary policy
  "cold turkey" vs. "gradualism," 181, 205–6
  cost of disinflation, 181, 200, 206
Downs, A., 136n.1, 142
Drazen, A., 109, 116, 133
Driffill, J., 172, 173
Dynamic inconsistency, 160, 173, 200

Economic conditions conducive to deficits, 131–32
Economic man, 5
Economic model, 11
Economic stimulation
  increased monetary growth and bias toward, 169
  through surprise creation, 162, 163, 193, 194
  trade-off between inflation and, 163, 193, 194
Economy, private, 112–13
Education, investments in, 133
Efficiency, criterion of, 5
Elections, 138
  cost of democracy and, 136
  with fully informed voters, 146
  with partially informed voters, 136, 142–45
  voters with diverse objectives, 137, 151–54
  welfare loss after, 136, 143–44, 145
Enelow, J. M., 60n.11
Equilibrium, voting, 127–28. *See also* General equilibrium effects
Equity, vertical, 55
Euler equations, 166, 167, 168, 197
Exchange, politics as, 5
Expectations of monetary growth
  average credibility and, 191–92, 199–200
  marginal credibility and, 192, 200
  rationality of, 159–60, 161, 164, 167–68, 182, 193, 195–98
  model of formation of, 193, 195–97; proof of, 167–68, 197–98
Expected value of governmental objectives, unconditional, 174–75, 186–87

Federal Open Market Committee (FOMC), 159, 162, 194
Federal Reserve, 158, 159, 181, 191, 193
Feldstein, M. S., 133
Fellner, W., 182n.32
Fiorina, M. P., 36
Fiscal policy, bequest-constrained individuals and, 110
Fischer, S., 137n.5, 161, 163n.7, 181, 206

Forecasting ability of government, 139; reelection and, 144–45, 146, 147, 153–54
Franchise, extension of, 23, 24, 25, 30, 32–33, 37, 46, 71
Free (liberal) society, 8, 9–10
Frey, B. S., 137n.5, 145n.16, 161
Friedman, M., 9, 17, 53, 54n.3, 146, 180, 192, 201, 204–5
Friedman, R., 53, 54n.3

Gale, D., 115n.8
General equilibrium effects
  characterization of political equilibrium in presence of, 127–28
  of debt financing, 122
  determination of debt in absence of, 122–23
  individual attitudes toward debt in presence of, 123–25
Goodfriend, M., 159, 181
Goodhart, C. A. E., 161
Gordon, D. B., 159, 160, 170–72, *189*, 192n.3, 205, *212*
Government. *See also* Debt
  authoritarian, 14–15
  decision rule for monetary growth, 166–68, 194, 197–98, 210–11
  discretionary policy and
    facing voters with diverse objectives, 137, 151–54, 155
    fully informed voters, 146
    partially informed voters, 136, 142–45
  flexibility of, 136, 141, 147
  forecasting ability of, 139; reelection and, 144–45, 146, 147, 153–54
  objectives of, optimal level of ambiguity and, 178, 179
  revenue
    cash-equivalent mix and, 73–74; per capita, 66–67
  services of, suppliers of, 36–37
  size of, increased, 16, 23–52
    data on, 48–51; decisive voter and, 25, 31, 32–33, 37, 40–42, 46, 70; economic environment and, 25–28; economic growth and, 33; equilibrium government policy and, 60; individual-collective relationship in decision making, 33; informed voter and, 33; majority rule determination of, 23; mean income and, 25, 31, 32–33, 46; productivity and, 29, 30, 31; reasons for, 23, 36–37; supply side model of, 36–37; voter demand and, 46–47
    voting rules and, 29, 30, 33; Wagner's law and, 31–32, 33, 37, 44–46, 47
  spending of, classification of, 38n.2, 49–50
Grants, 48
Green, E. J., 167
Growth, economic
  deficits and, 131
  government size and, 33

Haberler, G., 182n.32
Hague, D. C., 55
Harberger, A. C., 55
Hardouvelis, G., 172
Hayek, F. A., 8, 9, 24, 33
Helpman, E., 32n.11, 42n.8
Hetzel, R. L., 158, 163n.8, 193
Hinich, M. J., 60n.11
Hotelling, H., 25
Household, modification of standard theory of, 65
Huizinga, J., 137n.5, 161
Humphrey-Hawkins Act, 194

Ignorance, veil of, 5, 6
Impossibility theorem, 85
Incentives, 25
　in-kind redistribution and, 55
　negative income tax and, 53–54
　progressive tax structure and, 98
　redistribution and, 67
Income
　ability and, 82
　absolute, 32
　aggregate, 45
　distribution of
　　deficits and, 131; in developing countries, 14–15; progressive tax structure and, 96–97; tax rate and, 25
　earned, 27, 41
　equilibrium relation between taxes and, 39–40
　mean
　　government size and, 25, 31, 32–33, 46; ratio of median income to, 46–47; relative to decisive voter's income, 25, 31, 37; tax rate and, 31
　median, 45–47
　per capita, 28
　pretax, 26, 27–28, 38
　redistribution and, 17
　redistribution of, 10–11, 16n.11, 25
　relative, Wagner's law and, 32, 33, 44–45
　tax rates and, 17
Income tax, 71–72, 80. *See also* Negative income tax; Progressive income tax
Inconsistency, dynamic, 160, 173, 200
Individualism, methodological, 5
Individual productivity, 26
Inflation, policymaker's trade-off between economic stimulation and, 163, 193, 194
Inflationary bias, 169, 170
Inflationary objectives, 200
Information
　asymmetric, 142–45, 148, 155, 170, 171, 175
　public response to, in democratic society, 146
Initiative, 12–13
In-kind transfers, 18, 55, 56, 71
Interest rates

　change in welfare due to changes in, 124, 125–27
　debt-induced increase in, 129
　real, 112, 117–18
Intergenerational reallocation of consumption, 123–24, 125
Intertemporal reallocation of taxes via debt, 118, 119–20
Inter vivos bequests, 118n.9

Jaffe, D. M., 179, 182
Johnson, W. R., 16n.11, 26n.5
Judicial decisions, changing, 12

Kalven, H., Jr., 18, 76
Kane, E. J., 159n.1, 193
King, M. A., 26n.5, 31
Kleiman, E., 179, 182
Klevorick, A., 85
Koester, R., 18
Kormendi, R., 18
Kramer, G. H., 77, 85
Kuznets, S., 33
Kydland, F. E., 21, 135, 146, 171, 173, 192n.3, 200

Labor, tax rates and, 29
Labor-leisure choice, 79–83, 97
Labor productivity in production of public good, 148, 149
Labor supply, redistribution and, 64
Larkey, P. D., 24, 25n.4, 32, 36, 44
Legislative branch, 10
Leisure
　consumption and, 26, 38–39
　marginal utility of, 81
　as normal good, 26, 28, 39
　redistribution and, 58–59
　tax rates and, 29
Leisure-labor choice, 79–83, 97
Liberal (free) society, 8, 9–10
Lindbeck, A., 56n.7
Linear consumption tax, 54–55
Little, I. M. D., 55
Logrolling, 7
Logue, D. E., 160, 179, 182
Lombra, R., 159n.1
Longevity, deficits and, 131
Lucas, R. E., Jr., 136n.3, 163n.7

McCallum, B. T., 137n.5
MacRae, D., 137n.5
Majority rule, 9–10, 110, 118
　government debt and, 120–28
　government size and, 23
　income redistribution and, 10–11
　progressive income tax and, 78–79
　　collective intransitivities induced by, 85; nonworking decisive voter and, 99–100; tax schedule parameters and, 83; working decisive voter and, 95–99

restrictions on, 10
Marginal credibility, 192, 200
Market processes, 4, 13
Maximum principle, 6n.2
Meckling, W., 8
Median-income voter. *See* Decisive voter
Median-voter hypothesis, 32–33
Meltzer, A. H., 23, 25n.4, 37, 46, 54, 55, 68n.18, 70, 71n.21, 77, 80, 95, 103, 110n.1, 111, 116, 120n.12, 127n.22, 132, 137, 152n.26, 159n.1, 163n.7, 181n.31, 193, 194n.7, 202n.18, 206
Methodological individualism, 5
Meyer, L. H., 181
Mieszkowski, P. M., 53, 54
Milgrom, P., 144n.14
Military draft for defense, 148–50
Mirrlees, J. A., 55, 77
Mob rule, 37
Monetary announcements, 21, 191–213
 credibility of, 20
  determinants of, 199–201; measures of, 191–92, 199–200; model for analyzing, 193–97; uncertainty under alternative monetary arrangements and, 201–5
 effect of, 201–3
 legal requirement of, 191, 194
Monetary growth, 158–90
 credibility and determinants of distribution of, 159, 168–70
 derivation of government's decision rule for, 166–68, 194, 197–98, 210–11
 framework for analyzing choice of, 161–64: comparison with previous and current literature, 170–74
 monetary control quality and, 159, 160, 176, 181–82, 193–94, 200, 204, 205–6
 optimal predictor of, 165, 167, 182–85, 195–97, 206–10
 policymaker's objective function and, 161–64
 politically optimal level of ambiguity in policy, 158–59, 174–80, 181–82, 187–88
 public's beliefs and expected rate of, 164–65
 rationality of expectations of, 159–60, 161, 164, 167–68, 182, 193, 195–98
  model of formation of expectations, 193, 195–97; proof of, 167–68, 197–98
 variation of planned rates of, 160–61
Monetary regimes, 21
Monetary uncertainty, 169–70, 185–86
Money, changes in, 20–21
Money growth forecast error, variance of, 169–70, 185–86
Moran, M., 159n.1
Mueller, D. C., 24n.3
Musgrave, R. A., 6n.3, 24n.3
Muth, J. F., 165n.11

Negative income tax, 17–18, 53–75
 appeal of, 69–70
 Congress and, 54
 decisive voter's choice of, 60–64, 67–69, 71
 Friedman's case for, 53–54
 incentives to work and, 54
 model economy of, 56–60
Neo-Ricardian world, 109. *See also* Debt
Niskanen, W. A., 7, 22, 24, 33, 34, 36, 52
Noise in money control, 179, 182
Noll, R. G., 36, 52
Nominal-contracts framework, 181
Non-bequest-constrained individuals, 119, 120, 121, 128n.23
 bequest chosen by, for given structure of taxation and redistribution, 116–18
 debt determination in absence of general equilibrium effects, 122–23
 effects of debt on welfare of, 125–26
Nordhaus, W., 137n.5
Nozick, R., 3–4
Nutter, G. W., 3, 23, 36

Okner, B. A., 26n.5
Okun, A. M., 160, 179, 182
Olson, M., Jr., 24, 33
Operative bequest motive, 109–10, 119
Ordeshook, P. C., 24, 37n.1, 42n.8

Pareto-efficient frontier, 152–53
Participants, passive and active, 16. *See also* Voting; Decisive voter
Peacock, A. T., 23, 24, 32, 36
Pechman, J. A., 26n.5, 53, 54
Peltzman, S., 24, 36
Pensions, 19
Phelps, E. S., 77
Pigou, A. C., 77
Policy
 ambiguity and credibility about, 16
 discretionary action vs., 19–20
 inflationary bias of, 21
Policymakers. *See also* Monetary announcements; Monetary growth
 credibility of, 20–21
 reelection motives of, 20
Political economy
 use of term, 4
 welfare economics vs. positive, 15–16
Politically motivated government
 facing voters with diverse objectives, 137, 151–54, 155
 with fully informed voters, 146
 implementation of constitutional rules in, 147–48
 with partially informed voters, 136, 142–45
Political processes, 4, 13, 118–19
 individual preferences in intertemporal structure of taxation and, 118, 119–20
 redistribution through, 113
Politicians, motivation of, 135. *See also* Politically motivated government
Politics as exchange, 5

Pommerehne, W. W., 54n.3
Poor, bequests and the, 19
Porter, R. H., 167
Portfolios, debt and capital as substitutes in, 110
Prescott, E. C., 21, 135, 146, 171, 173, 192n.3, 200
Pressure groups, 7
Private economy, 112–13
Production, debt and capital as substitutes in, 110
Production function of public good, 149
Productivity
  decisive voter's choice and, 72
  government size and, 29, 30, 31
  individual, 26
  of labor in production of public good, 148, 149
  redistribution and, 58–59
  tax rate and, 29–30
Progressive income tax, 18–19, 76–108
  arguments for, 76
  decisive voter's choice about tax schedule, 83–95
    globally winning schedule, 91–95;
    majority-winning schedule, 85–91
  labor-leisure choice and, 79–83, 97
  majority rule and, 78–79
    collective intransitivities induced by, 85;
    nonworking decisive voter and, 99–100;
    tax schedule parameters and, 83; working decisive voter and, 95–99
  as optimal form, 76–77
  political process and government budget constraint, 83–91
  in presence of public good, 100–101
Public debt. See Debt
Public good, 30
  deficit financing and expenditures for, 132
  information as, 146
  production function of, 149
  productivity of labor in production of, 148, 149
  progressive income tax and, 100–101
Punishment mechanism, 172

Quadratic tax function, 18

Radner, R., 144n.15
Ramsey, F. P., 55
Rawls, J., 5, 6
Real interest rate, 112, 117–18
"Record of Policy Actions of the Federal Open Market Committee," 194
Redistribution, 7. See also Debt; Government; Tax(es)
  basic model of, 17
  characterization of individual economic decisions for given structure of, 114–15
  decisive voter's reasoning about, 70–71
  incentives to work and, 67
  of income, 10–11, 16n.11, 25
  income and, 17
  individual and collective decisions about, 16
  in-kind, 18, 55, 56, 71
  intergenerational, 19
  labor supply and, 64
  leisure and, 58–59
  median income and, 45–46
  political economy model of, 14
  through political process, 113
  productivity and, 58–59
  ratio of mean to median income and, 46–47
  tax rates and spending for, 77
  voter demand and, 46
  Wagner's law and, 46, 47
Reelection(s)
  forecasting ability of government and, 144–45, 146, 147, 153–54
  maximizing probability of, 151–54
  welfare and, 136, 142–45
Referendum, 12–13
Resourceful, evaluating, maximizing man (REMM) model, 8
Retired persons, 19, 30–31
Richard, S. F., 23, 25n.4, 37, 46, 54, 55, 68n.18, 70, 71n.21, 77, 80, 95, 110n.1, 111, 137, 152n.26
Riker, W. H., 24n.3
Roberts, J., 144n.14
Roberts, K. W. S., 25, 26n.5, 29, 37, 42, 60, 72, 77, 80, 95, 110n.1
Roe v. Wade, 12
Romer, T., 26n.5, 32n.11, 36, 42n.8, 55, 77, 80, 84, 85, 95, 110n.1
Rosenthal, H., 32n.11, 36, 42n.8
Rule(s). See also Voting rule
  active support vs. passive acceptance of, 11–12
  choosing and changing, 11–15
  constitutional, 6–7, 20, 137, 146–48, 154, 173
  derivation of government decision rule for monetary growth rate, 166–68, 194, 197–98, 210–11
  discretion vs., 19–20, 135
  for monetary growth, uncertainty and, 204–5

Sadka, E., 77, 82
Sahota, G. S., 31
Samuelson, P. A., 111
Sargent, T. J., 166
Schneider, F., 137n.5, 138, 145n.16, 161
Seiglie, C., 19n.14
Shepsle, K., 7
Sheshinski, E., 26n.5, 80
Simons, H., 77
Simultaneous-equations bias, 43
Snyder, J. M., 77
Social planner, problem of, 138–41
Social security, 30–31, 33, 37, 110, 118

Index

Social welfare. *See* Welfare, social
Stalin, J., 10
Stiglitz, J. E., 55
Stolp, C., 24, 25n.4, 32, 36, 44
Stone-Geary function, 38
Subsidies, 48
Suffrage, universal, 14, 29. *See also* Voting
Supply-side model of government size, 36–37
Supreme Court, 12
Surprise creation
   economic stimulation through, 162, 163, 193, 194
   politically optimal level of ambiguity and, 174–75, 176, 181

Tax(es). *See also* Negative income tax; Progressive income tax; Redistribution
   characterization of individual economic decisions for given structure of, 114–15
   consumption, 54–55
   direct, 55, 71n.20
   economic welfare analyses of, 7
   equilibrium relation between income and, 39–40
   income, 71–72, 80
   indirect, 55, 71n.20
   individual and collective decisions about, 16
   intertemporal structure of, 118, 119–20
   linear, 26
   political-economic processes influencing, 13–14
   redistribution and, 37
   social security benefits and, 37
Tax-induced distortions, 132–33
Tax possibility frontier (TPF), 84–85, 89
Tax rate, 26, 38–39
   average, 77–78
   effective, 77, 78
   extension of franchise and, 71
   income and, 17
   income distribution and, 25
   increased, effects of, 29
   marginal, 77–78, 80
   mean income and, 31
   productivity and, 29–30
   productivity of decisive voter and, 40–41
   spending for redistribution and, 77
   voting rule and, 25, 29–30
Tax schedules, 18, 79
   globally winning, 91–95
   majority-winning, 85–91
Taylor, J. B., 159, 160, 163n.7, 173n.21, 181, 205
Time preference of policymakers, 192, 204; ambiguity level and degree of, 180, 182
Tobin, J., 53, 54
Tocqueville, A. de, 24–25, 37
Transfers
   cash-equivalent, 54, 56
   economic welfare analyses of, 7
   in-kind, 18, 55, 56, 71

Transversality condition, 166n
Tullock, G., 6, 10n.5, 24n.3, 33

Uncertainty
   under alternative monetary arrangements, 201–5
   monetary, 169–70, 185–86
Unconditional expected value of governmental objectives, 174–75, 186–87
Unemployment, 206
U.S. Congressional Resolution, 133, 191, 194
Universal suffrage, 14, 29. *See also* Voting
Utility function, 26
   in private economy, 113
   Stone-Geary, 38

Varian, H. R., 86n.10
Variance of money growth forecast error, 169–70, 185–86
Veil of ignorance, 5, 6
Vellrath, M., 137n.5
Vertical equity, 55
Vietnam War protest, 12
Vote, extension of the, 23, 24, 25, 30, 32–33, 37, 46, 71
Voters
   decisive. *See* Decisive voter
   with diverse objectives, 137, 151–54, 155
   government size and demands of, 46–47
   informed
      fully 146; government size and, 33; partially, 136, 142–45
Voting, 3–4, 7–8; turnout and restrictions on, 42
Voting equilibrium, 72–73; characterization of, 127–28
Voting rule, 15
   decisive voter's choice and, 70, 71
   defined, 29
   government size and, 29, 30, 33
   income redistribution and, 25
   tax rate and, 25, 29–30

Wage rate
   average, 112
   change in welfare due to changes in, 124, 125–27
   deficits and, 131
   in private economy, 112
   proportion of bequest-constrained individuals in population and, 117
Wagner's law, 17, 31–32, 37
   interpretations of, 44–45
   ratio of mean to median income and, 47
   redistribution and, 46, 47
   relative income and, 32, 33, 44–45
Wars, deficits during, 132
Wealth
   deficits and, 131
   inherited, benefits of larger debt to, 129

Wealth *Cont.*
 proportion of bequest-constrained individuals in population and, 117
Webster, C., Jr., 181
Weil, P., 116
Weingast, B., 7
Weintraub, R., 158, 193
Welfare, social
 of bequest- vs. non-bequest-constrained individuals, effects of debt on, 125–27
 combined effect of debt financing on, 122
 crowding out of capital and, 110–11, 121–22
 loss after elections, 136, 143–44, 145
 maximization of, by social planner, 139–41
 politically motivated government and, 142–46
 probability of reelection and cumulative, 151–54
 total effect of change in level of debt on, 123–25
Welfare economics, 7, 15–16
Welfare payments, 71
Wicksell, K., 5, 11, 24
Willet, T. D., 160, 179, 182
Winer, M., 24, 25*n*.4, 32, 36, 44
Wiseman, J., 23, 32
Wolff, N., 19
Wooley, J. T., 158, 193